American Museum ᵒꜰ Natural History

POISON

SINISTER SPECIES WITH DEADLY CONSEQUENCES

MARK SIDDALL

Illustrations by Megan Gavin

Sterling Signature
NEW YORK

Sterling Signature
NEW YORK

An Imprint of Sterling Publishing
387 Park Avenue South
New York, NY 10016

Text © 2014 by Mark Siddall
Illustrations © 2014 by Megan Gavin

ISBN 978-1-4549-0764-0

Distributed in Canada by Sterling Publishing
$^{c}/_{o}$ Canadian Manda Group, 165 Dufferin Street
Toronto, Ontario, Canada M6K 3H6
Distributed in the United Kingdom by GMC Distribution Services
Castle Place, 166 High Street, Lewes, East Sussex, England BN7 1XU
Distributed in Australia by Capricorn Link (Australia) Pty. Ltd.
P.O. Box 704, Windsor, NSW 2756, Australia

Art direction by Chris Thompson
Book design by Yeon Kim

For information about custom editions, special sales, and
premium and corporate purchases, please contact Sterling Special
Sales at 800-805-5489 or specialsales@sterlingpublishing.com.

Manufactured in the United States of America

2 4 6 8 10 9 7 5 3 1

www.sterlingpublishing.com

4. Things That Bite
143

<< >>

INTRODUCTION

We tend to think that our world is, on the whole, a safe place. And on average it certainly is. The awkward thing about statistical distributions and averages is that some small portion of events always lies at the edge of expectations, in the tail of the distribution—the rare event. The likelihood that any individual reader of this book is going to be bitten by a venomous snake, stung by a box jelly, or poisoned by eating broiled fish is very small. But almost certainly, one among you will. Wading barefoot on certain beaches, fancying particular seafoods, even performing certain forms of worship may predispose a person to peril.

In the end, probabilities are useful for actuarial purposes, like for insurance companies and medical prognoses. But as the great philosopher and professor Karl Popper (1902–1994) noted, there are no probabilities than can be associated with singular rare events that may befall an individual at a specific point in time. All of which goes to say that there's no point in being afraid of deadly snakes, cone snails, scorpion fish, or stinging insects—not even the worst of the venomous ones—unless you make a point of touching them, in which case the probability of harm jumps very quickly to certainty.

What are poisons, toxins, and venoms? Venom is the easiest to define because whatever it is, it's delivered mechanically with a bite or a sting, with teeth, a spine, or another weapon. Today we tend to think of venomous animals as those one shouldn't touch or be bitten or stung by, and poisonous ones as those we shouldn't eat. If a poison is something delivered in food or drink or by inhalation, then there are no poisonous snakes, only venomous ones—you can safely eat any snake (but you shouldn't, as many are threatened with extinction).

What, then, is a toxin? The ancient Greeks gave us the word, related to archery: A *toxikon* was a poison arrow or really any poison that could be applied to something sharp, even a sword or a spear. Modern usage tends to refer to a toxin in terms of a specific molecule. A poison or a venom may be made up of dozens of toxins with different effects. Pit viper venom is a good example, with its muscle-liquefying myotoxins, nerve-wrecking neurotoxins, cell-killing cytotoxins, and hemo-toxins that incapacitate the blood's own clotting system. That is, toxins are more often defined by what they target than where they come from. A suitable example is tetrodotoxin. This powerful neurotoxin is present in the livers of pufferfish, like fugu; in the skin of the rough-skinned newts of Oregon; and in the bite of the blue-ringed octopus of the Indo-Pacific. In the latter, it's part of the octopus's venom; in the others, it's a poison—regardless, it's a neurotoxin capable of shutting down nerve impulses going to every muscle (except the heart).

The Greeks' word for a poison like the hemlock Socrates was forced to drink was *pharmakon*, the root for our modern pharmacology—revealing the dual relationship we have had with substances holding equal powers for ill and for health.

There are many books covering toxic animals, be they venomous or poisonous. Some are compendiums of critters; others are atlases of colorful beasts provided as much for their aesthetic beauty as for ease of recognition. A topic this compelling for its flirtations with fear, danger, and death easily lends itself to wide Internet coverage, much of which is tailored to top-ten lists and to our ever-increasing attention deficit. It is hoped that this book provides a more personal glimpse of the intersections of toxic animals with people. Human experience with toxic animals, stories of individual encounters, and the manner in which many have been inspired and others scarred are fascinating, tragic, and occasionally comedic—mostly tragic, of course, but the stories are no less compelling.

That the animals themselves are as compelling as are the stories is a tribute to the creativity of natural selection. Inasmuch as a noxious animal may better defend itself, advertising its perniciousness with striking colors and obvious patterns, it may not have to resort to any defense at all. Whether black, red, yellow, or green, each is easily recognizable, especially in combinations, and each is an evolved strategy to warn aggressors to reconsider. Is it merely coincidental that these are the most common colors on national flags? Often, too, it is enough to have the appearance of possessing weapons of mass destruction, even if there actually are none in the end. For nearly every truly toxic beast, there's a conniving mimic.

Müllerian mimics are like the new tough kid on the block, trying to look as mean as the most successful ruffian already on a street corner. Batesian mimics are all bark and no bite.

For every powerful offense, there's an equally resourceful defense. It's quite natural that we should misunderstand the true natural dynamic of a toxic world. Unseen snakes in the grass with flesh-eating venoms raise hairs on the backs of necks. The potential for painful scorpion stings as a result of slipping on an unchecked boot is an effrontery of the most personal sort. Yet from an evolutionary standpoint, we have very little to do with it, however often we might get in the way of an arms race. The most powerful poisons, the most virulent venoms, are held by those animals in greatest threat of being eaten by something that has been developing resistance. In the Red Queen's running game, it's not about getting ahead so much as not being left behind.

Another generality, evident in several toxic systems, is the principle of bioaccumulation. Frogs the world over, birds in New Guinea and Africa—possibly even the common quail—all appear to be capable of accumulating toxins from various elements of their diet. In some instances, the result is accidental; in others, it appears to be an evolved powerful deterrent to predation. Even tetrodotoxin, present in such a wide variety of animals, is produced by aquatic bacteria, not by the poisonous or venomous animals themselves. Raise any of these in captivity and they're rendered harmless—even the fabled Japanese roulette of overpriced fugu has been defanged by controlled aquaculture methods that carefully control diet. Bioaccumulation of natural toxins is usually more difficult to predict and manage. And though this book concerns itself with naturally occurring toxins, it's worth noting that the principles of bioaccumulation explain why, conservation priorities aside, eating whale meat is a profoundly stupid thing to do—the amount of mercury that has moved up the food chain is mind-numbing.

Meanwhile, our own species has been terrifically resourceful. What plants, anemones, corals, jellyfish, and toadstools all have in common is a relative inability to run away. And yet for these species to still exist, they must be able to defend themselves against the constant onslaught of predators and herbivores. Such defense is often accomplished with poison. For thousands of years, we've been harnessing the poisonous properties of plants. This is not limited to traditional medicine—the latest, greatest antimalarial drug is artemisinin from the sweet wormwood plant, much like quinine was from the cinchona before.

Less frequently, human cultures have leveraged the powers of animal toxins. South American and African peoples have tipped their darts and arrows with animal toxins for millennia. But animal toxins for medicinal purposes (excluding the psychedelic side) is a more recent evolution of human medicine. The first purified animal toxin used in medicine seems to have been Jeorg Haas' deployment of hirudin from leeches in the first human dialysis treatment in 1924. That presaged a century of discovery to the extent that we now have treatments for diabetes from Gila monster venom, painkilling peptides from cone snails, blood-pressure medications from pit viper venom, and possible treatments for multiple sclerosis from sea anemones. If there ever was a compelling reason to mobilize for the preservation of biodiversity, this is it! Never mind panda bears and other cute and cuddly mammals—it's the nasty snakes, the stinging insects, the blood-sucking beasts, and the toxic jellyfish that we must not lose.

With such a colorful, fearsome subject, the amount of hyperbole out there about some of the nasty animals described isn't surprising. It can be difficult to know what to believe sometimes. Occasionally, however, there is a kernel of truth. Black widow spiders actually do like dark enclosed places, where their insect prey might be readily attracted to odiferous gunk. So notwithstanding my earlier dismissal of probabilities—if, like 80 percent of black widow—bite victims in the 1800s, you're reading this in an outhouse . . .

◄◄ ►►

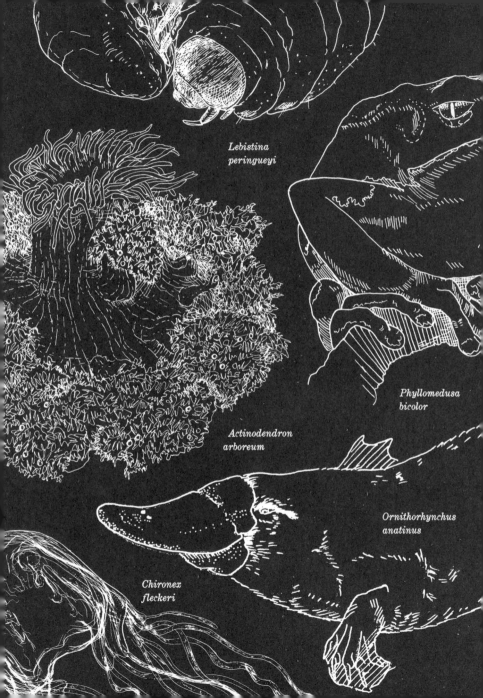

Lebistina
peringueyi

Phyllomedusa
bicolor

Actinodendron
arboreum

Ornithorhynchus
anatinus

Chironex
fleckeri

1

THINGS ONE SHOULDN'T TOUCH

Jousting Knights

COMMON NAME: Platypus

SCIENTIFIC NAME: *Ornithorhynchus anatinus*

SPECIES RANGE: Eastern Australia and Tasmania

Like the giraffe, *Ornithorhynchus anatinus*, the platypus, is one of those animals that appears to have been created by committee, with each member constructing a part independently that *should* have made for a successful beast. One designed the otterlike body, a second made the duck-billed mouth for snagging insects, the third fashioned a beaver's tail for storing fat, and a fourth knew that geese feet were best for swimming. Alas, no one was put in charge

of waste removal! This task must have fallen to the gynecologist, who just worked with what he already had. The single hole the platypus possesses for this is reflected by the name monotreme (*mono* for single and *treme* for hole). Among the monotremes are the platypus and the echidnas, each with a monotrematous condition that is shared with all vertebrates from fish on up—except for those mammals that evolved later. The "duck-billed" adjective is unnecessary, as the duck-billed platypus is the *only* platypus—just "platypus" will do. And while *platy* means flat, and *puss* is slang (from Gaelic) for mouth or face, the name platypus has nothing to do with its being duck-billed. It derives from the Greek word *platupous*, meaning flat-footed, which it is.

Unfortunately, the genus name *Platypus* was already taken by another flat-footed animal when the prolific George Shaw gave the odd animal its first taxonomic description in 1799. Known for his attention to detail, Shaw wasn't convinced that the specimen he had received was not an elaborate hoax perpetrated by a wily taxidermist. We owe the more appropriate "bird nose" name of *Ornithorhynchus* to the great English naturalist Sir Joseph Banks (1743–1820), who first honed his skills in Newfoundland, Canada. At the Grand Banks, Sir Joseph was studying penguins—the Great Auks in the genus *Pinguinus*, not the Antarctic penguins of the genus *Aptenodytes*—when he met the intrepid Captain Cook (see page 71). Their friendship flourished to the extent that Sir Joseph would later take the role of ship's naturalist on board the HMS *Endeavor* for the first of Cook's global circumnavigations. Naturally, Sir Joseph was tapped to go on the HMS *Resolution* in 1772 for the second of Cook's voyages. However, after making a fuss about his quarters on the *Resolution*, and even after the Admiralty spent over 10,000 pounds sterling to make it to his liking, Sir Joseph refused to go. Replacing a fuddy-duddy with a moaning bellyacher, Cook later chose Johann Forster in Sir Joseph's place. It was Forster who saw what we now call penguins in the Antarctic, and who nearly died from eating fugu (see page 72). It is ironic that Sir Joseph's platypus remains extant in its awkwardness, while the less awkward auk is extinct.

The 180-foot-wide Mount Hypipamee crater on the Atherton Tablelands of Australia has 240-foot-high sheer granite walls encircling a deep watery hole called a diatreme, which owes its origins to a volcanic blast that occurred some 100,000 years ago. Leaning precariously over the edge, I resolved that it would be quite impossible to look for leeches in this particular pond as there was no way in and no way out. Actually, there *was* an easy way in, as the lone platypus below had clearly discovered. This unlikely monotreme in a diatreme seemed perfectly happy wandering unmolested in the greenish waters. I don't know whether anyone eventually rescued the poor platypus. The prospects are precarious, as platypodes are poisonous— venomous, to be precise.

Sir Joseph Banks missed the half-inch-long venomous spur jutting out of the rear legs of his specimen. This overlooked oddity on male platypodes was, however, noted later with curiosity by Sir Everard Home, who is better known for describing the first Ichthyosaur discovered by Joseph and Mary Anning at Lyme Regis near their home. Sir Everard, working with a preserved platypus in England, thought the spurs functioned to restrain unwilling female platypodes *in copulo*.

In 1818, Sir John Jamison (1776–1844) discovered the painful truth when he accidentally stuck one such spur in the hand of a volunteer while helping him haul a male platypus from the water. As the very first person to be granted free title over land in Australia, Sir John would have been able to study these strange animals at will, while acquiring as much acreage as he could. Eventually amassing vast tracts and wealth, Sir John founded the Bank of New South Wales and led a lavish and hedonistic lifestyle right up to his passing in 1844. Sir John left us a detailed account of the platypus envenomation he had witnessed in his volunteer, describing the resulting swelling from the spur's point of inoculation: prodigious sweating, vomiting, and painful lockjaw. Ever the self-aggrandizer, Sir John took credit for saving the volunteer's life by applying salad dressing to the wound and dosing the victim with opium. Not long after, the lie was put to this exaggeration— local aborigines understood the stings to be painful and yet never fatal.

Sir Everard Home, however, could not be disabused of his erroneously erogenous coital conjecture, and doubled down with a claim that females have corresponding leg pockets for the receipt of the barbs. Unfortunately for Sir Everard's reputation, no one else could see this mythical pocket. Sir Richard Owen even found rudimentary spurs on a few female specimens. Still, Charles Darwin wrongly insisted that the spur was without venom—a forgivable error, as problematic stings were not a consistent eventuality of periodic platypus pokes. The matter was laid to rest by Sir John Jamison's antithesis in probity, the pious Reverend William Webb Spicer. Reverend William—who also started a bank, but one of the benevolent penny-savings variety—correctly determined the male platypus venom to be seasonal, associated only with the time of year that males combat for the affections of females, a bit like jousting knights. Unlike the many other be-knighted explorers above, Cook and Darwin each were without peer, and yet also without peerage.

<-<- ->->

Blue Persuasion

COMMON NAME: Greater blue-ringed octopus

SCIENTIFIC NAME: *Hapalochlaena lunulata*

SPECIES RANGE: Indo-Pacific Ocean

ow long he had been lying on the beach was probably no longer clear to the German diver. Although unable to move, he would, nonetheless, have been completely aware of his surroundings. He might have even still heard the waves breaking over the reef edge nearby as well as the sound of his own heart still beating. Fully cognizant of the light of day creeping in through partially closed eyelids, his lungs wouldn't have been filling with what air was still coming in ever-shallower breaths. Some vacation to the Andeman Sea that turned out to be. *"Dummkopf!"* he might have cursed to himself. Going into the *Wasser* alone was *verboten*; he knew that. He had picked up that little octopus. It looked friendly enough. It even seemed to be happy when it flashed those pretty little spots of blue. So what if he had perched it like a parrot on his shoulder

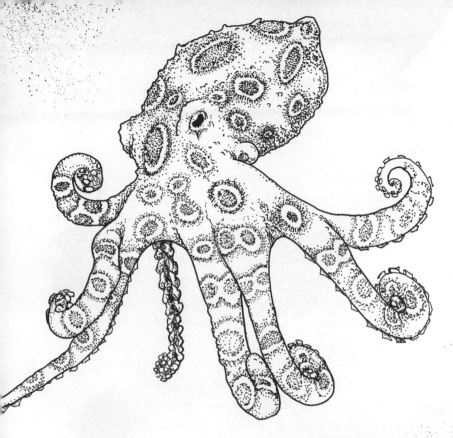

while walking out of the surf? He was going to put it back; really he was. "*Gott*, doesn't anyone walk this stretch of beach?" Unable to fill his lungs, darkness and unconsciousness finally overcame him and sadly he wasn't found—not in time.

While the greater blue-ringed octopus is small enough to place in the palm of a hand, this is just about the worst thing to do with any of the three or four *Hapalochlaena* species distributed about the Indo-Pacific region. Normally, their pale tan to brown bodies are as difficult to see as those of any other octopus on a coral reef. Octopodes are concealers, and change not only their color patterns to match their surroundings, but also their whole body shape and form at will. They can transform from smooth to knobby to lumpy in a heartbeat. Larger octopus species rely on the strength of their eight arms as well as a large, sharp beak to subdue and consume their prey, but these little blue-ringed cephalopods aren't so brutishly endowed. Instead, they pounce on crabs half their size, or even half-stranded fish much bigger than themselves, dealing the prey a deadly venomous blow. The toxins include a cocktail of neurologically active amines such as dopamine and acetylcholine, but in doses that are usually not problematic for humans. However, in addition to these lesser poisons, they also pack a tetrodotoxin punch.

Thankfully, most snorkelers and divers respect all critters that live on and around coral reefs with a mannerly and courteous look-but-don't-touch attitude. Each of the three or four cases of fatal bites from a blue-ringed octopus—and the many more nonfatal ones—have been violations of this sacrosanct code. In each instance, a mortified mollusk macerated its manhandler only after being unceremoniously lifted out of the water. Tetrodotoxin poisoning by injection bypasses the slow gastrointestinal process that the toxin faces after, say, eating fugu. Instead, symptoms from a bite include rather prompt zombification. The victim (or shall we say the perpetrator, actually) is quickly paralyzed; the arms and legs become floppy and nonresponsive. Eventually, the poison has a terminal effect on the muscles needed for breathing. This all occurs, however, without much in the way of any cognitive impairment. Recently, a child playing in the shallows of a beach near Cairns, Australia, suffered a bite on the hand from lifting an attractive octopus out of the water. Like others, he was able to survive after a day of intensive care and mechanical ventilation in a hospital as the toxin dissipated. The boy later reported having been fully conscious (and quite terrified) throughout the long ordeal.

Hapalochlaena species are great swimmers. And while they can be really quite stunning when they turn on those intensely blue iridescent rings of their mantle and arms, this display happens only when they have already seen you. The warning coloration is an indication that the octopus is terrified. It won't attack, but do give it its space. A little appreciative respect is all that's needed . . . or else.

<-<- ->->

Father of the Year

Without a doubt, the Big Bad Wolves of the sea are the box jellies. Unlike the loitering drifters that their scyphozoan jellyfish cousins are, cubozoan box jellies are directional swimmers with twenty-four eyes (all the better to see you with). The deadliest among them, the sea wasp (*Chironex fleckeri*), has four massive, stinging "hands" (all the better to hug you with), each with more than a dozen stinging tentacles.

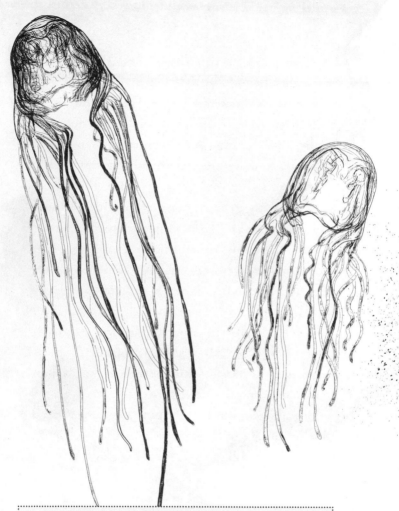

COMMON NAME: Sea wasp

SCIENTIFIC NAME: *Chironex fleckeri*

SPECIES RANGE: Indo-Pacific Ocean

A box jelly's fleeting life begins microscopically at the mouth of freshwater rivers, where it mixes with brackish marine water and settles on the bottom to pass the cool months of the austral winter and spring. As summer approaches, in October, and waters warm, tiny jellies arise in search of food. A voracious predator, the box jelly grows quickly to four pounds on a diet of ever-larger shrimp and fish. Regrettably for its scores of human victims, the box jelly's four tufts of ten-foot tentacles are merely diaphanous, easily torn tendrils. Evolved to avoid being torn apart by a thrashing shrimp desperate to escape a deadly embrace, the toxins injected by the cubozoan's stinging cells are sufficiently powerful to kill their prey instantly—and errant humans much more slowly and painfully.

From Thailand through Indonesia, the Philippines, and on into Australian waters, *Chironex fleckeri* and a few related species have been responsible for several thousand deaths in the last 120 years, about five times the number of fatalities as there have been from shark attacks. Tragically, children are the most likely to succumb to the neurotoxic and cardiotoxic venom, injected by microscopic box jelly nematocysts, which induces paralysis, asphyxia, and cardiac arrest within minutes of being strafed across bare skin. And while a healthy adult may not succumb so readily, even the burliest Aussie triathlete would have a considerable challenge swimming to shore while doubled over in indescribable agony.

A clever understanding of box jellies and their stings has done much to mitigate the likelihood of deadly encounters. With speeds of up to four miles per hour, box jellies are usually on the move beginning seven days after a full moon. Their gelatinous bodies are essentially transparent or ghostly white. But white is a shade they appear to ignore; they're not using their eyes to find one another. Dark objects are readily detected while red ones are avoided. Stinging nematocysts are too small to penetrate even the thinnest of fabrics, and are activated chemically, not by touch. They also will not fire if the pH of the water is too low. So the next time you are in Australia, don't be alarmed. That's not a cross-dressing sun worshipper in search of fish and chips on the beach wearing bright red nylons and carrying a bottle of vinegar, but rather a lifeguard on duty about a week after the first full moon of summer.

The sea wasp is not the only box jelly worth fretting about. Get stung by the minuscule Irukandji jellyfish (*Carukia barnesi*) and fretting is quite possibly all you will do for the next week: which is to say that while this cubozoan's sting is only sometimes fatal, no one will be able to convince you of that when you are facing Irukandji syndrome symptoms, including vomiting, a racing heart, and a powerful, overwhelming sense of impending doom. The connection between *Carukia barnesi* stings and Irukandji syndrome was demonstrated in 1961 by physician Jack Handyside Barnes, with a stunning disregard for any Father of the Year Award. Barnes was so certain that these malevolent jellies— each of which can have yard-long stinging tentacles trailing from their quarter-inch box-bell bodies— were responsible for Irukandji syndrome that he slapped a bunch of them on himself, on a lifeguard, and on his nine-year-old son. The ensuing frantic ride to hospital and twenty-four hours of intensive care indelicately proved his point.

--- ->-

Beetlemania

> COMMON NAME: Rove beetle
> SCIENTIFIC NAME: Staphylinidae family
> RANGE: Global

There's more to drought than lack of water. First the rains don't come; then the grasshoppers do. And as though a parched and ravaged landscape isn't enough, a plague of blister beetles (Meloidae) arrives to round out the drought's trifecta. The pattern has been repeated again and again, from the dustbowl of the 1930s to the drought of the late 1980s, which cost the modern equivalent of $100 billion. Terrible droughts have recently occurred in Mali in 2004, and more recently in 2010 and 2011, from West Texas to Idaho with the highest acreage designated "exceptional drought" in recent memory. As there's less and less to eat, grasshoppers congregate on what little remains, and the simple act of rubbing against each other turns them into swarming clouds of ravenous egg-laying locust. The eggs then feed the next year's beetles. For every adult blister beetle there were about twenty grasshopper eggs that didn't make it through the winter. This scourge of adult beetles can be just as destructive as locust because they are also quite fond of a major component of hay: alfalfa. Horses are known to have consumed fatal doses in their feedstock. Blister beetle toxins are very stable. Spraying pesticides can make things worse because the caustic remains of dead beetles just concentrate in place.

Bombardier beetles, in the family Carabidae, are capable of shooting hot liquid jets—though not enough to raise the kind of welts that blister beetles manage. Rove beetles are worse. A peculiar subsection of the family Staphylinidae, rove beetles are armed with paederin, a toxin as powerful as some cobras' venom (be thankful rove beetles are small). These conniving beetles don't produce paederin on their own: The blistering agent is manufactured by symbiotic bacteria bioconcentrated by beetles for battle. Some species of rove beetles are pretty plain looking, but, like most of the insect nasties, many are black and red for ease of recognition and avoidance. The genus *Paederus* comprises some six hundred species throughout the globe. In South America, they are variously known as *bicho de fuego* or *podó*. Population booms of "tomcat" rove beetles are problematic near Surabaya, West Java. With streamlined bodies that curve at the rear, they resemble fighter jets in profile, thus their Indonesian nickname, after the F-14 jet. In Iran—the only country to still deploy the U.S.-made F-14 Tomcat—there are periodic plagues of linear dermatitis, due to outbreaks of the rove species *Paederus fuscipes*. Several cases of widespread blistering raised concerns of chemical attack during Operation Enduring Freedom in Afghanistan (including one-tenth of the U.S. armed forces personnel at a single base) and later at the Balad Air Base in Iraq. In each instance, only blistering beetles were to blame.

Rove and blister beetles are generally harmless unless squished accidentally, or on purpose. Domino and bombardier beetles, on the other hand, can choose to deliver a well-aimed and directed blast. Domino beetles aim a searing shot of organic acid when sufficiently annoyed, which, by all accounts, is terrifically painful; but the number of accounts are few—there's even a thriving pet trade in this attractive *Anthia sexguttata*. Bombardier "beetlejuice" is a caustic concoction of chemicals kept separate inside the beetle's abdomen. When sprayed, these mix in an exothermic and corrosive stream that reaches the boiling temperature of water. It also makes a popping sound, serving to audibly reinforce the error of one's ways.

-<- ->-

Evolution Is for the Birds

COMMON NAME: Hooded pitohui

SCIENTIFIC NAME: *Pitohui dichrous*

SPECIES RANGE: New Guinea

Most evolutionary biologists have been asked countless times, "Why are there no . . . ?" Perhaps the most common kinds of questions are: "Why are there no marine insects?" and "Why are there no venomous birds?" It is not that these are uninteresting questions, but surely there are an infinite number of such questions that can be asked. Why are there no bloodsucking squid? After all, there's a vampire bat in the American tropics, a vampire bird in the Galapagos (the finch *Geospiza difficilis*), a vampire moth in Asia (*Calyptra thalictri*), a vampire fish in the Amazon (the candirú *Vandellia cirrhosa*), and even a vampire snail (*Cancellaria cooperi*) off the coast of Baja—but no vampire squid? Technically, *Vampyroteuthis infernalis* is the "vampire squid"; however, it spends its time feeding on detritic sea snow without ever relentlessly jamming a blood funnel into anything.

No one seems terribly bothered by the lack of flying monkeys, which is to say there's no cottage industry arm-waving about the lack of them. Why, then, does the lack of marine insects call out for some sort of deep evolutionary explanans? Functional biology explanations about osmoregulation do not hold much water— there are plenty of freshwater insects. Inevitably, someone will hypothesize something about "occupied niches" and a few eyes will glaze over. Honestly? The oceans are full? The real problem is that the question is not scientific because it's not testable. We should just accept the fact that sometimes it hasn't been "selected for" in the course of evolution—because, like flying monkeys, it simply has never happened.

Of course, sometimes the answer is even more simple: we haven't found it yet. Such was the case with the pitohui species of New Guinea—among the only known poisonous birds. California Academy of Sciences curator and department chair of Ornithology and Mammalogy, Jack Dumbacher, did what every parent tells their kid to not do: he licked his hands after handling a wild bird, without washing them first. The numbness and burning sensation in Dumbacher's mouth presaged his discovery of the only known instance of birds mounting a neurotoxic chemical defense against predation. The alkaloid homobatrachotoxin, which concentrates in pitohui skin and feathers, is identical to that found in the golden poison frog of Colombia, itself the most poisonous land animal known. The poison is an irreversibly binding sodium-channel neurotoxin with a steroid-based pentacyclic core skeleton, an intramolecular 3-hemiketal, and a seven-membered oxazapane ring—no simple molecule. It was thought impossible to synthesize in laboratories until 1998—just the sort of thing to get the intelligent designers going. How could this impossibly complex toxin be found in the skins of evolutionarily unrelated animals living 10,000 miles apart?

Common phenomena require common causal mechanisms and, in this case, it is soft-winged flower beetles in the family Melyridae, loaded with batrachotoxins and presumably relished by pitohui birds and phyllobatid frogs alike. Far from being a repudiation of evolution, the hooded pitohui's poisonous nature fits the expectations of natural selection perfectly. They advertise toxicity with aposematically bright colors, as well as a horrid smell, and are matched both by Müllerian and Batesian mimics (see page 51)—other pitohuis and the Greater Melampitta, respectively.

◄◄ ►►

Babar's Bar

COMMON NAME: Chrysomid beetle grub

SCIENTIFIC NAME: *Lebistina peringueyi*

SPECIES RANGE: Southern Africa

Amarula, a cream liqueur made from the fruit of the marula tree (*Sclerocarya birrea*), has made it to the global market, and is easily recognized by the drunken elephant on the label. Slightly cloying and tasting of caramel, this competitor of Irish Cream only faintly resembles the real deal. I once had the pleasure of polishing off a bottle of bootleg marula distillate after a long day mucking about in South Africa while trying to avoid being eaten by crocodiles. Fruity

and clear, the hooch packed a strong but sweet punch that left me regretting it in the morning.

The fruit of the marula tree, the "food of kings," has been used by the San and the Bantu peoples for over eleven thousand years. The macadamia-like nut and its delicious oleic acid–rich oils see enthusiastic culinary use. In addition, the gummy exudates from the bark of the marula tree have been used in making inks. The tree itself, which can reach over 50 feet in height, is an anacard (like poison ivy, mangos, and cashews), with enough toxicity in its rind to be used as an insecticide. However, it is the fruit's intoxicating value after fermentation that accounts for the marula's most coveted realizations. Mukumbi beer from above and below the Zambezi River is refreshing, but the stuff from Swaziland is what will put hair on your chest—if not also on your tongue. Interestingly, there were laws about carrying weapons during marula wine season in Namibia, where the fruit enjoys near religious reverence. And the stories about drunken elephants attracted from miles away to the smell of marula fruit fermenting on the ground are well known.

The notion that the marula's plum-size fruits drive elephants mad with drink once a year runs through both modern and historical travel literature. These facile claims—from nineteenth-century Zulu tales of loxodontine lushes "warming their brains," to modern marula bottle labels featuring plastered pachyderms—neglect two salient facts: First, elephants do not eat the fruit fermenting on the ground; they eat marula fruit right off the tree. Granted, a few fruit may fall to the ground after the tree gets knocked over, which bull elephants are want to do. Second, unless an elephant is unexpectedly lightweight, inebriating one would take over seven gallons of fully fermented marula juice, which is the unlikely equivalent of a truck full of about 1,500 fruit.

Nonetheless, there is no denying the "wasted elephant" stories. Experience shows that there is usually a kernel of verisimilitude behind something so widely believed, as there is in the case of the pie-eyed proboscideans of the low veldt. Along with eating the fruit of the marula tree, elephants are also known to strip off the bark. However, it is not entirely clear that the bark is what the behemoths are after. Underneath lurk the luscious larval grubs of the genus *Lebistina*. As a source of nutrition, beetle grubs are much richer in fats than the larvae of other insects like mopane caterpillars (see page 51). Rhinocerous beetle grubs grilled over a fire in Peru taste faintly of clarified butter. *Lebistina* species that live in marula trees are not recommended. They concentrate so much toxin in their bodies that the San people deployed them for millennia as an arrow poison, a fact well attested to in the journals of Finnish explorer Hendrik Jacob Wikar (1752–1818). Wikar deserted the Dutch East India Company and wandered among the San for several years. The value of Wikar's journals earned him a pardon, and they include a description of the San method for squeezing the guts of various grubs onto arrowheads. The slow action of the poison requires tracking an impaled wildebeast for several days, and perhaps 50 miles, before any gored gnu's a goner. Elephants—no lightweights, it seems—might just get drowsy and slightly sloshed from eating the grubs.

◄◄ ►►

All That Glisters

Although beautiful to behold, most brightly colored animals of a rain forest, especially those in plain view during daylight hours, are best not actually held—much less licked. Chuck Myers, a onetime curator at the American Museum of Natural History, knows this all too well. Some of his best-known work concerns research conducted in the 1970s on the poison dart frogs of South and Central America. Admitting that his assay method for finding toxic amphibians was efficient, "albeit not very elegant," Myers had taken to tasting frog skin.

COMMON NAME: Giant leaf frog

SCIENTIFIC NAME: *Phyllomedusa bicolor*

SPECIES RANGE: Amazon rain forests

Having seen a snake rendered incapacitated from biting a bright yellow-and-black-backed, blue-legged Golfodulcean poison frog in Costa Rica, Myers licked the frog. What ensued, by his understated account, was a numbing of the tongue, "followed by a disagreeable tightening sensation in the throat."

The Costa Rican frog, like other species in the genus *Phyllobates*, produces a variety of batrachotoxins in its skin. These simple alkaloids are among the most potent neurotoxic and cardiotoxic compounds known, orders of magnitude more poisonous than strychnine. There is no known antidote. Less than a decade later, in Colombia, Myers found a bright yellow frog with enough batrachotoxin in its skin to kill a hippo—a useful note, given that dangerous hippos are now invasive in Colombian rivers. After handling this golden poison frog, a species new to science, Myers noticed that the skin on his face burned for hours as a result of accidentally touching his beard. Wisely, this time he did not lick the frog. Had he done so, the result quite possibly would have been uncontrollable muscle contractions, heart palpitations, an inability to breathe, eventual asphyxia, and the consequential profound inability to describe the new species, rather aptly, as *Phyllobates terribilis*, meaning "terrifying leaf walker."

The Colombian golden poison frog remains unsurpassed as the most poisonous animal on the planet. An average adult frog has 200 to 500 micrograms of each of three kinds of deadly toxins concentrated in its skin: batrachotoxin, homobatrachotoxin, and batrachotoxinin-A. After witnessing the rapid death of mice exposed

COMMON NAME: Golden poison frog
SCIENTIFIC NAME: *Phyllobates terribilis*
SPECIES RANGE: Chocó rain forest, Colombia

to less than a microgram of skin extract, Myers quickly calculated that less than 200 micrograms would kill a human weighing 68 kilograms—a peculiarly precise, and perhaps self-referential, body weight to consider. Which is to say, one Colombian golden poison frog, if eaten, has enough toxin to kill more than a half dozen Chuck Myers.

While it is best not to lick poison frogs raised in captivity, this is entirely for the frog's sake. That is, if fed a diet of crickets, fruit flies, and mealworms, wild frogs are rendered harmless. Although it may take a few years, as in the case of the golden poison frog, this is also true of any other brightly colored South American poison dart frog species of *Phyllobates* and *Dendrobates*. Similarly, when any of these are raised from tadpoles in captivity—as is invariably done for popular exhibitions at museums and zoos—the frogs are as colorful and as nontoxic as a box of crayons. In the wild, brightly colored South American toxic frogs get their arsenal from the insects they naturally eat, which, naturally, does not include North American crickets. Toxic alkaloids from fire ant venoms are repurposed by and for the defense of many poison dart frogs in the genus *Dendrobates*, as well as by entirely unrelated African mantellid frogs. Batrachotoxin alkaloids in the skin of the golden poison frog, in contrast, are found in high concentrations in soft-winged flower beetles, as likely a food source for the frogs as they are believed to be for the only known poisonous birds living a world away in Papua (see page 37).

By way of clarification: regardless of the "poison dart frog" common names given to any of 175 species of red, blue, and green *Dendrobates* species, and notwithstanding many embellished tales of early encounters with Amazonian peoples, none are used for tipping blowgun darts. Only frogs in the genus *Phyllobates* have strong enough toxins for hunting, and this application appears to be unique to the Wounaan and Emberá peoples of Colombia, none of whom ever lived in the Amazon. Amazonian dart poisons are principally plant-based toxins, like curare. An advantage afforded to the supreme potency of *Phyllobates terribilis* is that it is spared the fate of the other two species of the genus used for tipping darts—these are both routinely tortured so as to induce enough poison. The Emberá only need wipe their darts on the bright yellow back of the golden poison frog before it is allowed to hop on its merry away.

◄◄─ ─►►

Advice from a Caterpillar

COMMON NAME: Puss caterpillar

SCIENTIFIC NAME: *Megalopyge opercularis*

SPECIES RANGE: Southern United States, Mexico, Central America

The underdogs of the insect world have got to be the caterpillars. These larval stages of moths and butterflies consume massive quantities of vegetable matter out in the open and during the day. Meanwhile, caterpillars cannot fly away (yet) and are speed-challenged by laughably, disproportionately small appendages that defy any notion of "intelligent" design. At best, a reliance on camouflage or the casting of chance to the wind on a thinly spun strand of silk can save a caterpillar from a predator, but this leaves life quite literally dangling by a thread.

The caterpillar Heimlich, from the movie *A Bug's Life*, deftly captures the downtrodden life of a lepidopteran larva during the circus troupe's first performance. As if to underscore his own inherited impotence, Heimlich enters the circus ring with an incongruous prosthetic black-and-yellow stinger applied to his rotund caboose, asserting to a disbelieving audience in his German accent, "I am a cute little bumblebee," anticipating the day that he will "be a beautiful butterfly, and zen everything vill be better." The climax sees Heimlich's role reduced to serving as live bait: "a tasty verm on a schtick." And yet caterpillars are hardly helpless. Most have risen above the challenges posed by their unwieldy shape and make up for their lack of speed by concentrating enough nastiness from the plants they eat so as to make themselves at least unpalatable, if not downright toxic. The best-studied examples are the Cinnabar moth (*Tyria*

jacobaeae) and the North American Monarch (*Danaus plexippus*), both of which advertise their presence with aposematic black-and-yellow stripes—like Heimlich's prosthesis. Conspicuously avoided by everything from birds to lizards, Cinnabars retain high levels of alkaloids obtained from their diet of ragwort. Monarchs have a heart-stopping beauty—quite literally. Cardenolide steroids and the cardiac glycosides retained from their diet of milkweed cause palpitations and congestive heart failure in any bird whose ancestors neglected to pass on an instinctual avoidance. The black-headed grosbeak (*Pheucticus melanocephalus*) is a notable exception and feasts on monarchs at will, as it should, for caterpillars are a rich source of protein. Mopane "worms," the larval form of the African moth (*Gonimbrasia belina*), are an important source of protein for millions of people from the Congo to Cape Town. Served raw, or fried alongside a healthy dollop of *nsima* (cornmeal) in Zambia, they are really quite pleasant, with a faint note of Earl Grey tea that is quite distinct from the split-pea characteristics of *beondegi* (silkworm pupae) in Korea.

To the extent that some species of adult butterflies draw attention to their inherent toxicity, others engage in explicit me-too-isms. Viceroys and monarchs are really quite difficult to tell apart, particularly in fluttered flight. They are equally toxic, too, taking mutual advantage of not requiring avian predators to have to think too much, or to remember too much, before deciding not to pluck them as a meal in midair. This Müllerian mimicry is outdone by the Batesian mimic pirates. Henry Walter Bates (1825–1892) went bamboo. Inspired by Darwin's *Voyage of the Beagle* (1839)—and long before goading Darwin into writing their earth-shattering theory of evolution by natural selection—Alfred Russel Wallace (1823–1913) and his drinking buddy Bates went on an entomological junket to the mouth of the Amazon. Four years later, Wallace, perpetually broke, decided to return to England on the brig *Helen* and sell his specimens to the highest bidder, which would allow him to pay off his debts. After twenty-five days at sea, the *Helen* caught fire. His entire collection went down with the ship, forcing Wallace to perpetually contemplate his perpetual penury for ten days as he drifted in an open boat in the middle of the Atlantic Ocean. Bates had remained behind in Brazil, and stayed in the dense jungle for a total of eleven years collecting about fifteen thousand specimens. When Bates finally returned to England and began sorting through an impossibly large collection of insects, he discovered that many were impossibly difficult to tell apart. Bates, who had no formal education, knew why. Many of the otherwise innocuous look-alike butterflies were mimicking the color patterns and shapes of toxic species. The subterfuge of the former species confers protection by virtue of predator avoidance of the latter.

In their own defense, caterpillars are hardly limited by toxicity and mimicry. The stiff spines of saddleback caterpillar moths (*Acharia stimulea*) and of io moths (*Automeris io*) pack a hefty punch that will quickly leave a huge welt that one can ponder while being overcome with waves of nausea for several days. Countless others deploy the finest of hairlike projections that produce hives. Impossibly thin, these irritating hairs will readily stick all over skin and in the eyes, and permeate the mouth and throat. Not uncommon in the southern United States, *Megalopyge opercularis,* or the puss moth caterpillar (one name among many common names), manages to combine both of these defenses into one. This Fabio of the insect world, the puss moth caterpillar has a luxurious mane of fine hairs that begs to be stroked as though it were a tiny Persian cat. They are best left alone, however, in order to escape the burning sensation that rushes up the arm and begins to blister, followed by headaches, chest pain, and considerable difficulty breathing.

It is the spines of South American species of the genus *Lonomia* that should *truly* be avoided on pain of death. Unfortunately, these Luciferian larvae are not particularly brightly colored and fail to provide warning of much consequence. Dr. Nikolai Ustinov, working in the Soviet bioweapons program, described the symptoms best. Ustinov accidentally got a needle stick during his weaponization research, infecting himself with the Marburg virus, which, like *Lonomia* and Ebola, causes disseminated intravascular coagulation (consumptive coagulopathy). He was moved to a bio-containment hospital with steel air-lock doors, and monitored by doctors wearing spacesuits. What began as bright red stars under his skin soon transformed into sweating blood from his pores, leaving unclotted smears across the pages of his journal while life slowly ebbed. When all blood-clotting factors activate at once, no clots are actually formed in any one place, which leaves the sufferer without any means to stop bleeding. Given that the entire body is constantly in the process of keeping blood from leaking unchecked—and that it can do so only with the aid of a healthy supply of those clotting factors— infection with hemorrhagic viruses like the Marburg virus or envenomation by the caterpillar spines of the giant silkworm moth (*Lonomia obliqua*)

renders one unable to stem the flood of blood—
anywhere. Unhappily, the caterpillar sting seems
to take about a week to run its fatal course.

⤛ ⤜

Asymmetrical Warfare

Anemones, corals, jellies, box jellies,
sea fans, and hydroids, together in
the phylum Cnidaria, are among the
most ancient forms of animal life equipped
with a responsive nervous system. They are also
equipped with specialized stinging cells called
cnidocytes, each of which has a nematocyst.
These are microscopic, highly coiled spring-
loaded weapons, often with backward-pointing
barbs. When triggered by touch or just the right
chemistry, like a whaling harpoon and cable let
loose from a cannon, the explosive discharge
forces the weapon into unsuspecting prey or

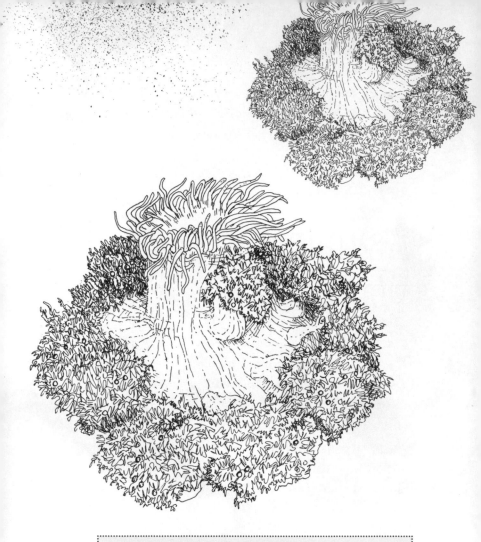

COMMON NAME: Hell's fire anemone

SCIENTIFIC NAME: *Actinodendron arboreum*

SPECIES RANGE: Indo-Pacific Ocean

offending foe. Once embedded, the neurotoxins follow. Most cnidarian toxins elicit only a mild reaction in humans, although there are notable exceptions, such as box jellies (see page 26), the Portuguese man-of-war (*Physalia physalis*, see page 126), fire corals, and a few anemones. Snorkeling in the Raja Ampat Islands, I was cautioned to give very wide berth to a hell's fire anemone (*Actinodendron arboreum*), which my guide ominously referred to as the "sledgehammer." The diaphanous pseudotentacles of antler anemones like *Lebrunia danae* are deceptively dangerous, too.

However, nudibranchs lie in wait. Neither Sun Tzu nor Clausewitz comment to much effect on one of the central tenets of modern asymmetrical warfare: abscond with your enemy's arsenal and use it to your own ends. Historically, this was simply ungentlemanly and lacking in chivalry. But the shell-less, sluggish gastropods called nudibranchs worked this out millions of years ago. Luscious-looking and ornamentally resplendent with patterns and color, the nudibranchs are masters of guerrilla warfare. It is difficult to describe their beauty and multitudinous form in words; try doing an Internet image search of "nudibranch"—trust me, you will be amazed. Were it not for the fact that nudibranch DNA points so decisively to a common ancestor with otherwise bland and nondescript gastropod molluscs, the nudibranchs collectively could have been mistaken for extraterrestrial invasive organisms. These colorful and ornate animals are bound to

COMMON NAME: Sea swallow

SCIENTIFIC NAME: *Glaucus atlanticus*

SPECIES RANGE: Oceans globally

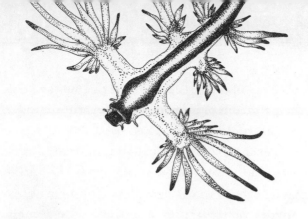

be advertising something. Nudibranchs rely more on their exquisite sense of smell than they do on sight. The delicious colors and patterning are not for others of their own kind, but rather, in most cases, they are advertising their own toxicity.

Well, not *exactly* their own. Nudibranchs eat cnidarians. *Spurilla neapolitana,* an aeolid nudibranch, for example, seeks out the antler anemones like *Lebrunia danae.* Of all of the nudibranchs, *Glaucus atlanticus* is perhaps the most decorative, with its vivid blue foliaceous body. This species dines extensively on Portuguese men-of-war, as it drifts upside down on the surface of the Sargasso Sea in the middle of the Atlantic Ocean.

Like any well-adapted seasoned insurgents, nudibranchs make deft use of the armaments they cannot create themselves but can nonetheless acquire by subterfuge. As they consume the flesh of their hydroid or man-of-war prey, the digestive juices of nudibranchs pass over the stinging cells. Perhaps this had origins in nudibranch ancestors avoiding being stung from the inside out, or perhaps not. Regardless, the nematocyst organelles of their prey are allowed to remain intact. What's more, they are actively transported from the slug's gut and deposited in the flowing fields of fingerlike projections on their dorsal surface. With a high-contrast color pattern denoting the location of the stolen arsenal, the venomous neurotoxic stinging apparatus of the nudibranch's own prey is thus deftly deployed in these tiny animals' own defense.

⤙ ⤚

Pincushions

COMMON NAME: Fire urchin

SCIENTIFIC NAME: *Asthenosoma ijimai*

SPECIES RANGE: Indo-Pacific Ocean

When looking for clues as to which animals might be problematic when touched or brushed against, it is often useful to pay attention to some common patterns: Bright colors, for example—especially contrasting ones—are the most obvious. But, there is the question of spikes—lots and lots of big, huge spikes. Sea urchins in the genus *Diadema* possess some of the longest spikes of all echinoderms. Although they are beautiful to behold, they are also venomous. Interestingly, it is their short spikes that are venomous, not the long ones. Many victims have complained of being envenomated by the long, black spines of these urchins, but not one of these accounts is likely to be true. To be envenomated by a species of *Diadema*, the long spines would have had to have

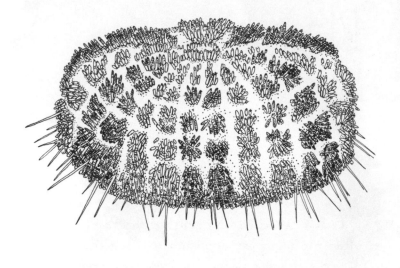

gone all of the way through the foot before the underlying spines could exert their venom. The long spines would break off in the skin before this could happen. The piercing is nasty enough, for sure, and prone to infection by noxious bacteria.

As if having to look out for long, black sea urchin spikes weren't enough, another thing to watch for when scuba diving is something terribly amiss relative to expectations. Urchins in the genus *Toxopneustes* have no prominent spikes at all. Instead, the three-pronged pedicillariae covering their hardened surface (the shell is called a *test*) pack a wallop while also making the urchin look like a little oceanic hydrangea inflorescence. When triggered, the jaws snap shut on skin and deliver both pain and a slight local paralysis. Stay away from the urchins that lack a hard test altogether! The pliable fire urchin (*Asthenosoma ijimai*) defends itself with neither a hard shell nor long, sharp spines. Therefore, it is of no surprise that this tam-o'shanter of an urchin is covered in painfully venomous spines. The venoms of urchins and other echinoderms are not deadly in themselves; still, screaming in pain while fifty feet underwater with a regulator in your mouth can lead to drowning.

⤙ ⤚

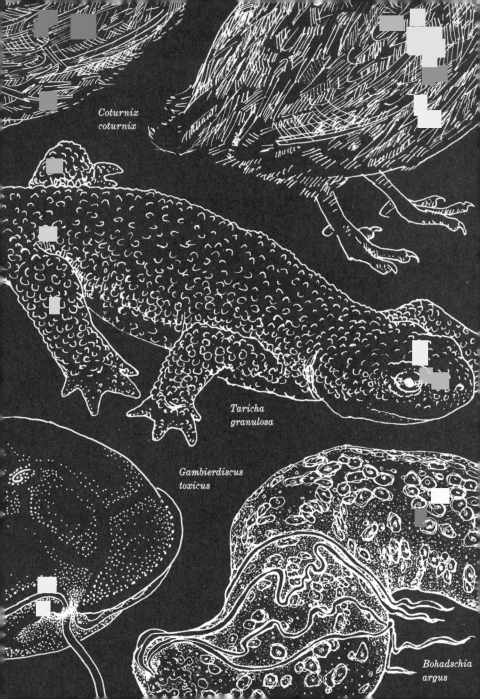

*Coturnix
coturnix*

*Taricha
granulosa*

*Gambierdiscus
toxicus*

*Bohadschia
argus*

2

THINGS ONE SHOULDN'T EAT

Fowl Passages

COMMON NAME: Common quail
....
SCIENTIFIC NAME: *Coturnix coturnix*
....
SPECIES RANGE: Europe, Asia, Africa

Unable to endure the unbearable feeling of nausea and the deep pain in her legs, a Greek retiree checks into the local hospital. A teenager enters the nearby clinic after fighting chest pains for more than twenty-four hours. A panicked Peloponnesian forty-something arrives at a hospital after feeling an excruciating ache in his shoulders and having been unable to keep any food down for three days. For all of them, standard urine tests proved anything but routine, since what flowed forth was neither clear nor yellow, but, instead, dark and chocolate brown! Tests would later diagnose each patient with rhabdomyolysis—a condition in which muscle cells dump proteins into the bloodstream, which the kidneys must then clear. Their muscles were liquefying.

Hebrew legend tells how thousands of years

ago, and already for two years, the Israelites had been eating nothing but manna. Granted, something that tastes a lot like wafers made with honey can't be all that bad, but dietary diversity is important, too. A total lack of animal protein can lead to pernicious anemia or to beriberi. Looking at chapter 11 of the Book of Numbers, we can forgive the more carnivorous of the recently emancipated desert wanderers for craving a little diversity. The loudest complained, "If only we had meat to eat . . . we have lost our appetite; we never see anything but this manna!" Begrudgingly, it seems, the Lord blew a mighty wind and drove up all of the quail from the sea. Plump little quail, stacked more than three feet high around the entire camp—rotisserie time! Alas, their god seems to have changed his mind about the ungrateful bellyachers, and "while the flesh was yet between their teeth, ere it was chewed, the wrath of the Lord was kindledt . . . and the Lord smote the people."

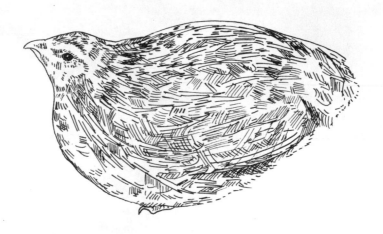

Symptoms of coturnism, a form of acute rhabdomyolysis in which inflamed muscle cells leak out their contents, can range from muscle pain to kidney failure. Similar crash-syndrome effects occur with deep crushing injuries to muscles as a result of car accidents. Coturnism, however, follows from eating European migratory quail (*Coturnix coturnix*), principally in the Eastern Mediterranean regions from Greece to the Red Sea. While the proximal cause of coturnism is a recent meal of quail, an earthly explanation remains as obscure today as it was during the Exodus.

Interestingly, it only takes an odd quail here or there to turn eating a delicious dish of Quail Baharat into a deadly game of Anatolian Roulette—the vast majority of the birds are perfectly safe to eat. The as-of-yet-unproven supposition that the birds are bioaccumulating some toxin goes back a long way. In fact, the Roman physician and philosopher Sextus Empiricus (c. 160 to c. 210 CE) assumed that noxious quail had eaten poisonous hemlock. But hemlock proved as fatal to quail as it was to Socrates. The Geoponica, a Roman equivalent of our Farmers' Almanac, blames hens hell-bent on the hellebore plant for coturnism. Whatever the cause, death-by-quail is geographically and temporally circumscribed. In the eastern range of the species, quail are toxic only during their autumnal southward migration; however, the seasonal incidence of coturnism is reversed during the migratory flight from Algeria to France in the westernmost range. That cooking the foul fowl is not effective prevention implicates a heat-stable toxin in the birds' largely vegetarian diet, very likely an ovatoxin-producing algae in the genus *Ostreopsis*. These marine palytoxins cause not only the characteristic rhabdomyolytic symptoms, but have also been responsible for recent and substantial Mediterranean poison outbreaks in 2005 and 2006. Understandably, and after leaving Kibroth Hattaavah (the graves of craving), the Israelites made do with manna for another thirty-eight years.

◄◄ ►►

One Fish, Toothish

COMMON NAME: Silverstripe blaasop

SCIENTIFIC NAME: *Lagocephalus sceleratus*

SPECIES RANGE: Indo-Pacific and Indian Oceans;
invasive in Mediterranean Sea

After leaving the New Hebrides (today Vanuatu), which the explorer, navigator, and cartographer Captain James Cook (1728–1779) claimed for England, the HMS *Resolution* sailed to the south so as to arrive in New Caledonia in time for some surreptitious eclipse watching. Careful observations of the moon's transit in 1766 determined the position of Newfoundland. Longitude was terrifically

important to navigation. As there would not be another eclipse in England until .1925, more remote eclipse encounters were needed. For a variety of cultures, a solar eclipse was not to be taken lightly. Watching the sun being completely swallowed up by the moon seems to have been terrifying to many. The fact that half of the people watching would have gone blind could only validate any belief of eclipses as portents of evil. Cook (who ate albatrosses) was not a superstitious man. There is no indication of foreboding evident in his journals before the South Pacific eclipse of September 6, 1774.

Returning from shore to the ship the following day and being quite famished, Cook, along with naturalists Johann Forster (1729–1798) and his son Georg (1754–1794), probably relished the thought of digging into an unusual-looking fish— the pufferfish *Lagocephalus sceleratus*—which the crew had acquired from the locals in exchange for cloth. The ships' illustrator was busy drawing the fish, and refused to allow its being fried up for dinner. But the fish had been cleaned and so the liver and roe sacs were fair game. After a single bite, the trio surely must have thought better of it. The burning sensation in their mouths and throats would have been immediate, and it is clear that there was no mystery about this fish being "off." By the wee hours of the morning and suffering the full effects of pufferfish poisoning, their skin felt as though it were in a blast furnace while simultaneously frosted over with ice, Cook later wrote. He was oddly incapable of distinguishing heavy objects from light ones, something that probably concerned Cook less than his constant and involuntary voiding at both ends as progressive paralysis set in. All of this from a very slight and mild exposure to what is now known as tetrodotoxin, a nerve poison that concentrates in the livers and reproductive organs of pufferfish.

Although the crew tried, no one could catch the ship's dogs after the poor hounds wolfed down the scraps. Luckily the dogs voided the offending liver on their own and survived, but a hungry pig that got into the trash didn't make it through the next day. On that following day, local Micronesians, seeing a pufferfish hanging on deck, frantically pointed and made dramatic gestures to the ship's crew which were, no doubt, the Micronesian equivalent of drawing a finger across one's neck.

Now, if Cook had visited Japan during one of his Pacific voyages, he would have known better than to eat the livers of these deadly fish. The Japanese have prized pufferfish for a very long time—their bones are found scattered among those of humans from the Jomon period (though it's unclear if there was a causal relationship). Two thousand years later, during the Warring States period, the fish bit back. Typhoons twice spared Japan from Mongol invasions in the thirteenth century. These kamikaze "divine winds" overturned most of the hastily built Mongol ships and sent the surviving crews limping back to Korean shores. The tables were turned in 1592, when all samurai were summoned to Hizen Nagoya Castle in Kyushu, Japan, to rally for a full-scale invasion of Korea. In a case similar to Cook having failed to consult the locals, a whole regiment of samurai were felled from having one last big fish-fry the night before embarkation. The Japanese invasions of 1592 and 1598 ultimately failed, and were disastrous for both countries in terms of casualties and treasure. One wonders whether the whole mess could have been avoided if the Koreans had invited the samurai ashore for an ill-considered dinner of fish liver in 1591. Instead, Japanese warrior-general Toyotomi Hideyoshi (1537–1598) went ahead and banned the consumption of fugu (the Japanese word for pufferfish, meaning "river pig"). Samurai would lose the entirety of their inheritances if they were caught slurping fugu from a bone. The ban allegedly was overturned three hundred years later by Japan's first prime minister, Hirobumu Ito (1841–1909), who apparently took a personal liking to it—and the deaths quickly resumed.

Postwar Japan under American occupation was not a pleasant place to be. An entire way of life was in shambles, the emperor lost his divinity, infrastructure and industry were obliterated, and citizens were starving everywhere, especially in the cities. Since people had no choice but to eat just about anything they could get their hands on, it didn't take long for entrepreneurial hibachi-ists to grill whatever scraps they managed to dig out of the trash bins from local restaurants. The death toll from street-grilled fugu entrails continued to mount until General Douglas MacArthur (1880–1964) created a strict fugu-licensing program that required safe preparation and thorough incineration of toxic pufferfish offal.

For some time now, the only deaths from fugu have been self-inflicted. The muscle meat of tetraodontid fish—named as such because they have beaklike buck-toothed jaws and include the species pufferfish, globefish, porcupine fish, boxfish, and blowfish—is perfectly safe to eat. Presently, close to ten thousand tons of fugu are eaten in Japan every year, about half of which is the farm-raised species torafugu, or tiger blowfish (*Takifugu rubripes*). Its consumption is more common in the southern regions— there are some eighty thousand fugu chefs in Osaka alone. All fugu chefs are licensed by a government agency, but only after demonstrating mastery of the proper cleaning, preparation, and disposal of pufferfish. The common-heard notion that there is a numbing sensation from fugu sashimi is entirely imagined by the patron— perhaps from too much soju. Any numbing sensation one experiences should be quickly followed by hospitalization. What's more, fugu is bland. Nonetheless, there is apparently a large culinary niche that offers $400 plates of tasteless fish artfully arranged in the shape of a chrysanthemum—which is, incidentally, a flower associated with funerals stretching from Krakow, Poland, to Kyushu, Japan.

Redfish

Cook's three voyages around the world are the stuff of legend and exploit, not to mention cartographic excellence. Arguably, his second Pacific voyage was probably his least pleasant. Cook eventually met his demise at the hands of a bunch of insulted Hawaiians on his third voyage. But it was during the second voyage, in which he had to put up with the

COMMON NAME: Dinoflagellate

SCIENTIFIC NAME: *Gambierdiscus toxicus*

SPECIES RANGE: Oceans globally

incessant whining of naturalist Johann Forster, that he fell ill, twice—once with "bilious colick," and then with a blocked colon—ran headlong into Antarctic ice, twice; and nearly died from eating fish, twice!

Several months before the New Caledonian eclipse "picnic"—the one that nearly extinguished Cook and the two Forsters with fugu liver—there had been a prior near miss with a fish—one that remains unidentified as it was quickly consumed by a hungry crew. Landing in the tiny and isolated island nation today known as Vanuatu, which Cook dubbed the "New Hebrides" (he was fond of antipodial Scottish place names that would annoy the French), an "away team" was briefly put ashore to meet the locals.

Cook was meticulous with his crew, keeping them on a rigid diet to keep scurvy at bay. One can imagine a salivating crew dropping anchor in azure blue, cool seas resplendent with reef fish. Absent Cook's stern gaze while he was ashore, the prospect of fresh food roasting on deck would have been a tantalizing and welcome respite from the daily doses of "antiscorbutic" sauerkraut and spruce "beer" rations. Yet, for a crew of 118, they'd need a lot of fish, or at least some big ones. By the time Cook returned to the ship, two huge, reddish-colored fish had been snagged and hauled on board. They were promptly cooked and divvied among the crew, with the officers and petty officers getting more than their fair share, of course. Vegetable accompaniment side dishes are not noted in Cook's journal. What is noted is that in "the Evening every one who had eat of these fish were seiz'd with Violent pains in the head and Limbs, so as to be unable to stand, together with a kind of Scorching heat all over the skin." The symptoms of scorching heat foreshadow the same neurotoxic symptoms Cook and the Forsters would experience three months later. According to ship surgeon William Anderson's account, at least five of the men suffered for four days before getting back on their feet. This time, unfortunately, the scavenging animals on board fared worse from eating the scraps, with the final death toll amounting to one dog, a hog, and a parakeet.

Ciguatera poisoning was well known to the Carib people of the Lesser Antilles, who attributed the condition to the cigua, a local turban snail (*Cittarium pica*) still prized today for stews in the Virgin Islands. Ciguatera is characterized by paralysis of the legs, uncontrolled salivation, and a burning sensation all over. It is interesting to speculate as to whether the crew of the HMS *Resolution* experienced ciguatera's characteristic dyspareunia, a peculiar priapism that is perhaps best not fully described in these pages. The poison responsible for causing such pain, ciguatoxin, would later be determined to be produced not by the kind of fish that Cook's crew ate nor by the marine snails that those fish might themselves favor, but, instead, by a microscopic single-celled algae, *Gambierdiscus toxicus*.

Like the algae that are responsible for toxic red tides, this little critter ebbs and flows with cyclical oceanic patterns. The problem is bioaccumulation. One little fish eats algae and absorbs some ciguatoxin. When a bigger fish eats ten little fish, it retains all of their ciguatoxin. Then huge fish eat the big fish, which ate the little fish, accumulating all of the ciguatoxin the little fish ate. So it stands to reason that the biggest and the most predatory fish are going to have the most ciguatoxin and will be the most dangerous to eat, like the case of the big red (probably coral trout) fish the crew ate on the *Resolution* on July 23, 1774—or like the odd death here and there recently from eating barracuda in the Florida Keys and Saint Thomas.

Depending on the dose and the individual survivor, symptoms of ciguatera poisoning can recur for over a decade, and can happen anywhere. However, it seems to be concentrated around small, isolated oceanic islands and atolls. The reasons for this remain obscure. And unlike annual warm-water red tide–related toxins, ciguatera is correlated with the cool phases of El Niño and the Pacific Decadal Oscillation. Too bad for Cook's crew, then, that Vanuatu is one of the most isolated reef systems on the planet, sporting even today a correspondingly enormous problem with ciguatera—especially from eating those very big, reddish reef fish like coral trout and emperor fish. It is also interesting to note that it was during one of the strongest Pacific Decadal Oscillations on record that the HMS *Resolution* plied the Pacific—the Hawaiians who killed Cook were "lucky" to even meet him.

-<- ->-

Bluefish

COMMON NAME: Atlantic mackerel

SCIENTIFIC NAME: *Scomber scombrus*

SPECIES RANGE: North Atlantic Ocean

As a direct result of Captain James Cook's three voyages, each of which included stops in New Zealand, Europeans were soon frequenting the Tasman Sea and the Southern Ocean, hunting whales and establishing a foothold in Māori Land. The Australia Company, in particular, had several large barques, each of which would be loaded in Leith, Scotland (old Caledonia as opposed to New Caledonia), with hardware, merchandise, and munitions to supply the growing British presence in Australia. Each would return home full to the gunwales with sperm oil. One of them, the HMS *Triton*, was captained by James Crear (1787–1859) following his exit from the Royal Navy. Crear would eventually settle in New Zealand for good and at a time when it was an unruly outpost that may well have outdone the lawlessness of Deadwood, South Dakota, that would follow some fifty years later. Crear seems to have been a man of considerable character, having once saved a fellow captain who was drowning, and even having taken a principled position against the "transportation" (read: "extraordinary rendition") of British prisoners to the South Pacific colonies. With the trade routes and shorelines masterfully mapped out by Cook, the *Triton* could make two return trips a year to Tasmania and New Zealand. During one such return trip, on April 17, 1828, some six hundred miles west of the Azores and more than one thousand miles east of Bermuda, calamity struck.

That morning, five members of the *Triton*'s crew, who had until then been feeling just fine, were acutely seized with remarkable, if distressing, symptoms, including violent headaches, capillaries nearly bursting from their bloodshot eyes, and bright red skin. Most alarmingly, their faces and whole bodies were so swollen with edema as to render them unrecognizable to their crewmates. The ship's doctor, Patrick Henderson, sprang to their aid, adding insult to injury. All five received powdered morning-glory root with the immediate, unhelpful effect of uncontrollable projectile vomiting, and two were given a good, long bleeding to relieve their headaches—an approach that could have succeeded insofar as it would have aggravated their deep shock to the point of passing out. The source of this sudden illness was no more a mystery to those present than was Cook's fugu poisoning fifty-four years earlier. All of the afflicted had breakfasted on the same skipjack tuna that morning. It is notable that Henderson attributed the sorry state of the fish to having been left overnight "while the moon was up," when in fact the real clue lies in his description that for "a number of days previous . . . everyone had partaken freely of the same fish without suffering the slightest inconvenience." It is fair to point out here that the miasmatic theory of disease would not give way to the notion of contagion for another half century; it would take another twenty-six years for John Snow to steal the handle off the Broad Street water pump.

Certain bacteria are able to extract energy from protein instead of sugar by ripping a carbon dioxide off of amino acids. In fact, the smell of some cheese and that unmistakable "fishy" smell of, well, fish, is a result of this breakdown to amines. Glutamic acid is broken down to glutamine; tryptophan is broken down to tryptamine; and a few select bacteria, like the lactobacilli in sauerkraut and yogurt, are able to break down histidine into histamine. Antihistamines had yet to be invented, so what the five crew members of the *Triton* were suffering from was acute anaphylactic shock, just as severe as if they had all been allergic to peanuts and had eaten Fluffernutters for breakfast. In this case, though, instead of their own immune systems dumping throat-choking levels of histamine into their systems, they received near lethal doses of histamine from fish that had been, by all accounts, lying about on board for some "number of days"!

Scombroid poisoning is probably the most common illness associated with eating fish. It is also the easiest to avert. Scombrid fish include various tuna, mackerel, bonito, and wahoo. That scythe of a tail fin attached to a narrow end is characteristic of fish built for speed. These fish have evolved a way to deal with lactic acid buildup in muscles that should be the envy of any Olympic sprinter: high concentrations of histidine serve as a natural buffer, keeping the pH near neutral. At these concentrations, though, the histidine is just too tempting for bacteria if the fish is mishandled or left at room temperature—the histamine by-product is toxic at a measly 50 milligrams per 100 grams of fish.

Alas, scombroid poisoning is hardly confined to those species, a fact this author knows from a personal experience with a bluefish. Bluefish, *Pomatomus saltatrix,* is one of my summer-time favorites. As an oily fish, it is perfect for grilling; just a little salt and some lemon tame this precious piscine pièce de résistance. My first inclination that I was in trouble should have been that nagging feeling in the back of my head that I may have somehow way over-spiced the fish when, in fact, I hadn't spiced them at all. My next inclination was total disorientation, wheezing, and considerable homage to a porcelain god.

It is difficult to get a handle on the precise

number of cases of scombroid poisoning, and this is a good thing. The only poisonings no one tracks are the ones that are rarely fatal, and this is one of them—knock wood, I'm not giving up bluefish. The connection between histamine poisoning and improperly handled fish (mea culpa) wasn't made until the 1940s. Most cases are reported from the United Kingdom, the United States, and Japan. Investigations following a 2008 outbreak of scombroid poisoning in Alaska found histamine levels in mahi-mahi that were easily three times the known toxic levels, and a case in the UK involving tuna-fish sandwiches sitting on a shelf all day before they were sold to unwitting victims had more than fifty times the allowable levels. Still, it is easy enough to avoid scombroid poisoning: make your own tuna-fish sandwich, and, if you didn't order spicy tuna, ask yourself why it feels so hot in your mouth.

◄◄ ►►

In a Pickle Briny

COMMON NAME: Leopard sea cucumber;
leopardfish trepang

SCIENTIFIC NAME: *Bohadschia argus*

SPECIES RANGE: Indo-Pacific and Indian Oceans

Fifty-three-year-old Abundio Golbe and his two sons had bagged some fat, juicy *balatan* for dinner Easter weekend in 2009, much as they had so many times before, as had countless Filipino generations before them. Complaining of stomach cramps and of profound thirst on Saturday, Golbe tragically did not make it to Easter Sunday dinner. Seven others were also hospitalized though thankfully spared Golbe's fate, four of whom were in critical condition for days. Something must have gone wrong in the preparation of the Golbes' fried *balatan*, as sea cucumber poisoning is exceedingly rare.

Sea cucumbers have been a delicacy in Asian diets for millennia. Most languages in the region have a word for the dish, such as *iriko* and

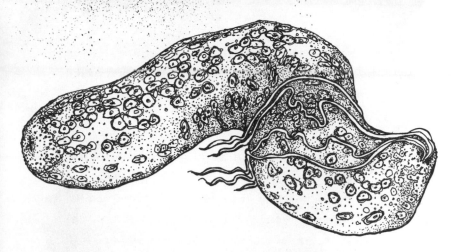

namako in Japanese, the Malaysian *gamat,* and the Filipino *balatan.* The Chinese *hai shen,* or "sea ginseng," is perhaps the most reminiscent of sea cucumbers' purported medicinal benefits. A century before Europeans encountered Australia's aboriginal peoples, Indonesian Makassars were visiting the northern coasts and trading largely for the purposes of garnering sea cucumber (trepang) fishing rights. With each monsoon season, first came rain and then came hundreds of trepangers. We now know that it was an established hunger for sea cucumber that first brought bolts of cloth, metal implements, rice, and even booze to the indigenous peoples of Arnhem Land. The European vegetable of the same name arrived much later. Presently, tens of thousands of tons of sea cucumber are harvested annually, worth tens of millions of dollars on the global market.

Related to sea stars and sea urchins, sea cucumbers are armed with holothurin toxins. These glycoside compounds are concentrated internally within tubular structures called Cuvierian organs. When poked, bitten, or otherwise insulted, many sea cucumbers will blast these tubules out of their bodies and into the surrounding water, releasing their toxins and— with any luck—slapping their sticky Cuvierian coils onto the eyes, gills, or other sensitive bits of offending predators. The effect is impressive. Holothurins, like other glycosides, are saponins, which is to say that for a fish, hanging around an irritated sea cucumber is much like swimming in concentrated dish detergent. Just a few ounces of holothurin per gallon of seawater will knock off every fish in a home aquarium. It should be no surprise, then, that sea cucumbers have long been used for fishing by cultures throughout the Indo-Pacific—unsporting, for sure, but highly effective in a closed lagoon.

For culinary adventurers, be aware that proper preparation of sea cucumber includes the removal of the inedible Cuvierian organs, followed by a thorough cooking of the meat. And, although sea cucumbers are among the most unassuming and the least worrisome critters encountered when snorkeling and diving, should one surreptitiously induce a cuke's ejecta, one would be well advised to *not* rub one's eyes in disbelief—there are a few anecdotal reports of blindness.

Blistering Bell–Bottomed Balderdash

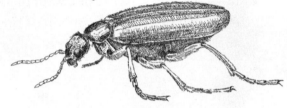

COMMON NAME:	Spanish fly
SCIENTIFIC NAME:	*Lytta vesicatoria*
SPECIES RANGE:	Europe and temperate Asian to Siberia

"Oh my, what have you done?" I was asked with mild alarm. "You're going to regret that," added the dermatologist, handing me some antibacterial ointment and a wad of gauze. Within the first two months of daycare my one-year-old daughter had experienced rotavirus, coxsackie, and now molluscum: tiny wartlike raised patches caused by a pox virus. We were told we could treat the spots with cantharidin "using a cotton swab stick," which would cause a blister under the pox and kill it. I resolved to first test it on myself. As the good doctor made herself busy, I spread a generous layer of the clear liquid over a 2-inch oval on the inside of my left forearm. Horrified, the doctor winced in empathy and explained that I had applied about a hundred times too much. It didn't hurt yet, but it would.

Cantharidin is a nasty terpenoid blistering agent produced by insects in the beetle family Meleoidae to protect their eggs, and themselves, from predation. These aptly named "blister beetles" have a global distribution. Many boast striking color patterns of black on red or bright yellows, which are variously displayed in polka dots, stripes, or more abstract patterns. One historically notable species, *Lytta vesicatoria*, the Spanish Fly, has a smooth, metallic appearance that ranges from copper orange to iridescent green. And, it is not a fly at all; it is very much a beetle of the cantharidian sort.

The aphrodisiacal properties of cantharides

are legendary, and legendarily balderdash. Their use dates to the classical Greeks, the ancient Romans, and even as far early and East as the Shang Dynasty. When consumed in certain quantities and concoctions, cantharidin is eliminated by the kidneys and urine. The result is a predictably "localized," shall we say, inflammation that might be mistaken to be "stimulation." Its purported effect of priapism in men has long, and erroneously, been presumed to lend the recipient to wanton satyrism and lasciviousness, regardless of gender. From Roman Emperor Augustus's wife, Livia Drusilla (aka Julia Augusta, BCE 50– 29 CE), through King Henry IV (1367–1413), and on to King Louis XIV (1638–1715)—which is, to say, the beetles of the Sun King, not the *Sun King* of the Beatles—Spanish Fly has been deployed for its lusty inducements. Italian physician Paolo Giovio (1483–1552) asserted that ground blister beetles were part of the Borgias' secret "la cantarella" poison. Although the Borgias never wrote their recipes down, accidental and criminal consumption of cantharidin speak to the dangerous Therapeutic Index, which is an index of the ratio of a lethal dose to an effective dose.

When they are too close together, the value of the index is far too low for safety. The Therapeutic Index of ethanol is about 10:1; for Spanish Fly it varies from 2:1 to 1:1. Death from cantharidin overdose is both very slow and very painful, and includes vomiting, thirst, inflammation of the stomach and kidneys, and considerable discharge of blood from the urethra and other orifices.

The Marquis de Sade (1740–1814) wasn't the first to fall afoul of the law, when, in 1772, he dosed a bevy of bacchanalists with baneful bonbons, leaving one Marseillaise of ill repute to suffer greatly; nor was he the last to be brought up on related charges. About a century later, Hoosiers Martin Bechtelheimer and William Young were sentenced to life in prison for the murder of an unwilling Susan Ingram. With forced doses of cantharides, instead of her affection and surrender, they managed to elicit only her untimely death, A lesser manslaughter sentence was awarded to the idiot Arthur Ford, manager of a manufacturing chemist firm near Tooting, in South London, who perpetrated the worst case of workplace harassment on April 26, 1954. Ford accidentally killed his two assistants (nineteen-year-old "Physical Excellence Girl" pageant winner June Malins and twenty-seven-year-old Betty Grant) with massive doses of Spanish Fly that he had hidden in coconut snow cones. Although they did not notice anything amiss at first, both victims were vomiting "nearly pure blood" within minutes, and both, in spite of intensive care and massive intervention by the heroic staff of St. James Hospital, were dead by the next afternoon. Arthur Ford had apparently learned of the supposed aphrodisiac use of blister beetles during his Second World War service in the British army.

Soldiering is often an unexpectedly dangerous business beyond the obvious ducking out of the way of bullets. This became obvious for "members" of the French forces of North Africa in the late nineteenth century, the situation "arising" on two separate occasions. One supposes that it mattered little to the French that *cuisses de grenouille* are not normally considered halal by Algerians. In both events, feasting on African frogs' legs resulted in widespread priapism that one attending physician described as "douloureuses et prolongées." This was, it seems, not unlike a regiment swallowing

a battalion's worth of Viagra. Like the golden poison frogs of Colombia (see page 45), frogs in North Africa are adept at concentrating the beetle toxins—dozens of which were found in the stomachs of the uneaten ranid remains.

Be careful what you eat. Like the hooded pitohui of New Guinea (see page 35), African spur-winged geese (*Plectropterus gambensis*) of the Gambia also concentrate beetle juice. A cantharidin-rich diet of meleoid beetles renders any such Gambian *foie gras d'oie* equally lethal to predator as to gourmand.

◄◄─ ─►►

Hold My Beer and Watch This!

COMMON NAME: Rough-skinned newt

SCIENTIFIC NAME: *Taricha granulosa*

SPECIES RANGE: Northern California to Alaska

The temnospondyls, which pre-date their modern amphibian descendants in the fossil record by about a hundred million years, were well equipped to defend themselves in terrestrial habitats. Variously armored with multitudinous sharp teeth, scales, and bony plates, many were enormous in size, reaching up to thirty feet in length and appearing superficially more like crocodiles than salamanders, newts, and frogs. The skin of modern lissamphibians (literally "smooth amphibians") allows them to dissipate carbon dioxide and absorb oxygen from their environment when their skin is moist. This can be particularly handy when stuck in

the mud waiting for the open waters of spring. To be efficient, though, amphibian skin needs to be exceedingly thin, offering only marginal protection against predators or from human industrialization.

It is no use being thin-skinned without also having mechanisms to avoid taking quick offense. Amphibians would not have been so globally successful without some of them producing enormous numbers of eggs in each clutch, others having expert camouflage, while still others deploy masterful toxic concoctions in their skin. Several newts have evolved with amazing defenses. The Spanish ribbed newt (*Pleurodeles waltl*) and the emperor newt (*Tylototriton shanjing*) of Yunnan, China, have toxic glands on the surface of their skin and sharp, bony ribs. If too much external

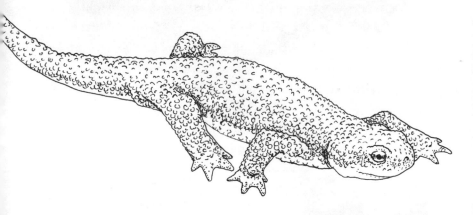

force is applied, as in the case of being bitten, the ribs puncture through the skin from the inside out and become slathered with dermal poisons. The sharp tips of these protruding ribs then inject the glands' contents into the offending predator.

However, natural selection doesn't have to play out in such an immediate manner. An adaptation, like skin toxicity, is a trump card when paired with obvious color patterns and displays. All that is needed is some brightly colored toxic newts to be ever so slightly less likely to be eaten than their drab relatives and for some common predators to be ever so slightly less likely to survive if they eat colorful prey. Over time, the predators that instinctively avoid orange newts, or spotted newts, will have longer lives and will produce more orange-and-spotted-newt—avoiding offspring. Over time, the brightly colored newts will live longer. Mimicry often comes into play. The North American medicinal leeches in the genus *Macrobdella* have bright orange bellies and distinctive orange polka dots along their backs—sufficient facsimiles of the central newt (*Notophthalmus viridescens louisianensis*) to avoid predation by gullible ducks and geese.

The rough-skinned newt (*Taricha granulosa*) of Oregon is relatively nondescript. From above, its warty skin is variously olive brown to drab gray. When startled or threatened, though, *Taricha granulosa* curls its tail skyward and arches its neck to reveal striking yellow to deep orange warning colors from underneath. It is not bluffing. The rough-skinned newt has over 3 milligrams of tetrodotoxin in every gram of skin tissue. This is the same deadly poison that makes fugu and blue-ringed octopodes so dangerous. Like a lot of over-the-top toxicity in the natural world, this one seems to be a consequence of garter snakes in the region having developed increasing resistance to tetrodotoxin. This Red Queen's game has been compounded by additional players hitting off the dealer. Aquatic caddisfly larvae, which love a breakfast of newt eggs, have similarly been developing resistance to the toxin with which female newts imbue their egg clutches.

There is an unsubstantiated story that relates how three Oregonian lumberjacks were found dead in the woods around an expired campfire. On the grill was a blackened coffeepot with a rough-skinned newt inside. The newt's body contains tetrodotoxin, which is remarkable for its stability to high heat. Boiled pufferfish is just as lethal, so it is confusing as to why anyone would intentionally eat such a tasteless fish.

Binge drinking with Oregonians can be hazardous. Heavily inebriated pub patrons have more than once discovered a lesson in taxonomy: rough-skinned newts aren't even closely related to goldfish. A thirty-six-year-old downed five newts during a night of drinking, which ended in hospitalization and being kept alive on a ventilator. He was luckier than the twenty-nine-year-old, who, after successive rounds of downing whiskey in a drinking game, accepted a deadly dare from his companions. Soon after swallowing his pet newt, he died of cardiac arrest. There are several remarkable elements to this true, if tragic, story. The most obvious is that there is enough tetrodotoxin in a single *Taricha granulosa* to kill a grown man; more remarkable is that people take pet newts to bars . . . and eat them.

Generally docile, stoic, and colorful, newts *do* make nice pets. The tetrodotoxin seems to come from bacteria in their environment, much like the skin toxins of South American frogs that are acquired from a diet of beetles and ants. However, even after more than a year in captivity eating an artificial diet, the probability of a rough-skinned newt remaining deadly toxic is extremely variable. So, please, do not eat your pets.

<+< +>>

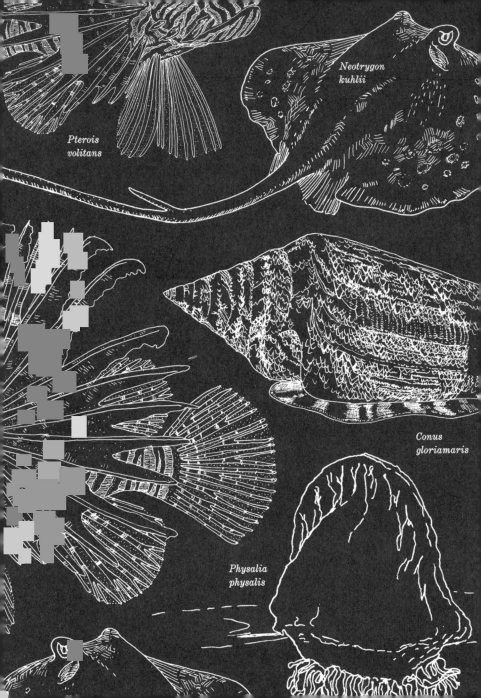

Pterois volitans

Neotrygon kuhlii

Conus gloriamaris

Physalia physalis

3

THINGS THAT
STING

Stinging Criticism

COMMON NAME: Arizona bark scorpion
SCIENTIFIC NAME: *Centruroides sculpturatus*
SPECIES RANGE: Sonoran Desert

Encounters with scorpions can be unnerving, whether or not one is stung. These seemingly headless eight-legged arachnids with prominent pedipalps (pincers) and menacing aculeus (stinger) mounted on a segmented tail look as though they were dreamt up by surrealist painter H. R. Giger, perhaps to be used as the next Space Jockey for another installment in the *Alien* movie franchise. Indeed, scorpions have undergone considerable exploitation on both the silver and the small screen alike, most of which get it horribly wrong. The "arachnids" of *Starship Troopers* have four legs and a head. The unlikely bright red scorpion in *Honey, I Shrunk the Kids* is at least believably nocturnal, an improvement over the *Clash of the Titans*'s rendition. As if the prospect of a scorpion crawling silently over skin isn't alarming enough, Hollywood has a penchant for putting them down pants. John Cusack needn't have been so distressed by Hilary Duff's character tucking away an Emperor scorpion (*Pandinus imperator*) in *War, Inc.*, since scorpions that rely on their stings for defense do not have the huge pincers seen in that scene. Do, however, avoid the piddling scorps in the family Buthidae, with pathetic pedipalps like the deathstalkers (*Leiurus quinquestriatus*) of North Africa crawling down character Ensign Stewart's trousers in *Dr. Phibes Rises Again*. Not once in the movie is it explained why the hoard of deathstalkers was curiously attracted to the victim's body. Vincent Price did not douse his victim with pheromones. Perhaps the restraining chair was juddering at 486 hertz? Either would work, but in both scenarios the scorpions would be trying to mate, not trying to sting.

There *are* some real cases of this scorpions-lurking-in-trousers trope. Ecologist Matt Leslie, returned from a week in Etosha National Park, Namibia, and, after donning his pajama bottoms, found one "just north" of where one wouldn't want to. In his book *Dinosaurs of the Flaming Cliffs*,

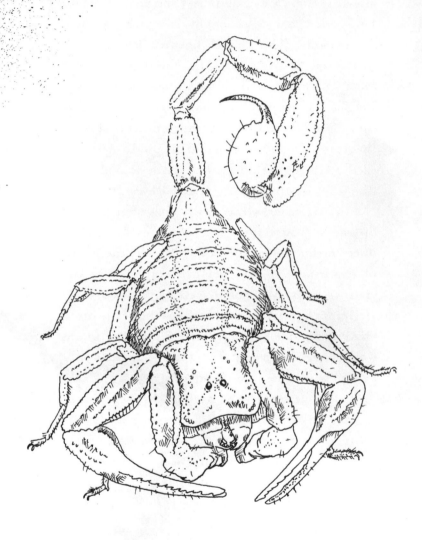

American Museum of Natural History paleontologist Mike Novacek relates how he once awoke to discover a pair of bark scorpions assaulting his groin. Insofar as this happened in Baja, California, the antagonizing arachnids were probably the relatively benign *Centruroides exilicauda,* which are not nearly as dangerous as their relative, the Arizona bark scorpion (*Centruroides sculpturatus*).

In the movie *The Crying Game*, Forest Whitaker's character relates the fable of the Scorpion and the Frog, in which a scorpion is endeavoring to find a way to ford a river—even the Chinese swimming scorpion (*Lychas mucronatus*) doesn't actually swim. The scorpion proceeds to ask the frog to carry him across the river. Being no fool, the frog rejects the notion on account of scorpions' supposedly notorious habit of stinging unprovoked. The scorpion then reasons with the frog that both would drown if it stung the astute amphibian, and the frog relents. Halfway across the river, receiving a mortal prick from his passenger, the frog asks, "Why did you sting me, Mr. Scorpion, for now we both will drown!" "It's in my nature," replies the scorpion. This cautionary fable, assigning blame to the victim for his naïveté about wholly unredeemable people, is an ancient one that is often repeated. In the Sanskrit original the victim is a turtle. In *Star Trek Voyager* episode "Species 8472," the scorpion is an allegory for Seven of Nine, and yet the victim in the parable is a fox. Forest Whitaker's accent isn't the least believable part of his telling of the Scorpion and the Frog fable: The notion of antagonistic scorpions stinging larger animals entirely unprovoked precisely inverts the dynamic. Scorpions with feeble pincers will use their neurotoxic venoms to immobilize their insect prey—it is a great way to avoid having their legs ripped off in the struggle. The searing intensity of the *Centruroides sculpturatus* sting, however, relates mainly to its being victimized by grasshopper mice in the genus *Onychomys*. These otherwise seemingly cute and cuddly rodents terrorize Arizona bark scorpions and feed on them with impunity. Indeed, if stung, the mouse repurposes the scorpion venom as an analgesic! In a dramatic validation of evolutionary biologist Leigh Van Valen's Red Queen hypothesis, bark scorpions have been evolving ever stronger and ever more painful toxins that bind to peripheral nerve cells, while grasshopper mice have been evolving ever greater lack of sensitivity to those toxins. Here, you see, "it takes all the running you can do, to keep in the same place." In this case, a scorpion's sting is better understood "through the looking glass" of defense, not offense.

Still, a scorpion sting can prove more than just a pain. Thousands of envenomations occur every year in the American Southwest and in Mexico. Thankfully, fatalities are rare. Very young children and elderly victims are the most likely to succumb to ensuing cardiac and respiratory arrest. Healthy adults, if stung, are best advised to wait out the pain, unless other symptoms arise, like shortness of breath or a thready pulse, in which case seeking medical attention prior to unconsciousness is recommended. Usually, treatment is palliative, though Ecuadorian missionary doctor Ron Guderian believed electroshocking the wound might help. When paleontologist R. Joe Brandon applied a hundred thousand volts from a stun gun to his friend Pat's scorpion sting, the current shot through both of them. The pain of Pat's sting was less noticeable relative to the pain of electrocution. Arizonan Marcie Edmonds was stung twice, and in quick succession. The first time was by a bark scorpion, which resulted in her throat closing, vision blurring, and muscles involuntarily tightening. The second sting was the $83,000 bill she got from her healthcare provider for two doses of antivenom.

Underexploited by Hollywood is the true story of the defenders of Hatra in 198 CE near present-day Iraq. Like Katniss of *The Hunger Games* forced up a tree, the Hatrans were outnumbered and under siege from Roman forces. Wanting for ammunition, Hatra's Parthian defenders are said to have scoured the sands at night for deadly deathstalker scorpions. In the light of day, the fruits of their nocturnal labors, packed tight in clay pots, were flung from trebuchets at their foes. Like the Career Tributes run off by Katniss' nest of tracker jackers, the Romans retreated both from the deadly stings and from abject terror!

<-<- ->->

Taken to Heart

COMMON NAME: Blue spotted stingray

SCIENTIFIC NAME: *Neotrygon kuhlii*

SPECIES RANGE: Indo-Pacific Ocean

lthough one can swim with stingrays in Belize and the Cayman Islands where it's possible to interact with and even to pet these ancient, graceful cartilaginous critters, one would be well advised to give stingrays a wide berth in places where they are not so well accustomed to the attentions of tourists. An alarmed ray will assume a defensive posture, arching its tail

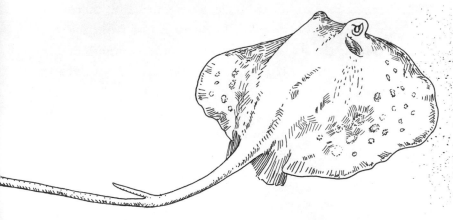

while exposing a sharp and venomous spine with serrations of backward-pointing barbs. Stepping on a ray is bound to have an unpleasant result, so shuffling your feet is a better way to get around in shallow waters frequented by stingrays. This goes equally for many freshwater regions of the tropics. Colombian health authorities report over two thousand injuries in any given year from potamotrygonid freshwater rays. There are at least three toxins bathing a stingray's spines, two of which are necrotic flesh-flaying poisons; the other is the same pain-inducing serotonin found in wasp venom (see page 132).

It is not surprising that the relative lethality of a stingray's defensive retribution is entirely related to where on the victim's body the spine has pierced. If it is an ankle, lower leg, or hand, there is going to be an awful lot of pain just from the trauma if not also from the serotonin component of the venom. If attended to, the wound should eventually heal, leaving a scar and a story to tell. If left alone, it will only take a little bit of time for the stingray venom to gradually liquefy the surrounding flesh, resulting in the real risk of these wounds: secondary bacterial infections.

Other anatomical locations of stingray spine penetration are considerably more traumatic, as "Crocodile Hunter" Steve Irwin's almost predictable fate illustrated.

In 2006, while filming the television series *Ocean's Deadliest*, Irwin got too close to a stingray. By some accounts, the ray was doing what rays do best, hiding just below the surface of the sand. Suggesting the encounter was anything but accidental, the official report notes that Irwin and his crew were actually filming when Irwin drifted over the ray. The standard stingray defensive posture ensued, and its barbed spine went through Irwin's chest, tearing into his heart. When Irwin reflexively pulled the barb out, the serrations would have done even more damage. Media accounts uniformly noted how "rare" it is that this kind of "freak accident" occurs; however, in a similar "rare" and "freak" accident in Florida only a few weeks after Irwin's untimely demise, eighty-one-year-old James Bertakis was trying to get a stingray out of his boat when it thrust its foot-long spine into his heart. Unlike Irwin, Bertakis resisted the instinctual yank at the spike, and it was successfully removed by surgeons. He survived. A little research suggests this may not be so rare. Five years earlier, a thirty-three-year-old man snorkeling at Coogee Beach in Sydney, Australia, was struck in the heart by a stingray, too. In this case, the ray took off with its spine intact. Rescued by lifeguards, the victim arrived in shock at the hospital with a racing pulse. The flowing blood from his lacerated coronary artery seemed to have helped him by washing away the venom. He was discharged six days later. He was luckier than a twelve-year-old boy who, in a situation that "occurred under freak circumstances" in 1989, was also struck in the heart by a stingray, in Queensland, Australia. Over a similar six-day time frame, venom proteins that were not flushed away caused the boy's heart tissue to fail, even though the right ventricle had been successfully repaired. Furthermore there is the reported "unusual bathing accident" suffered by a teenage girl in Hauraki Gulf, New Zealand, in 1939, and another incident in 1945 involving an Australian soldier enjoying a little R and R at St. Kilda, Victoria. The teenager received an eight-inch stab through her chest, inflicting a clean slice through her right ventricle; the soldier took it in the left ventricle six years later, with only a small, inch-wide wound visible on his skin. Neither survived. It should be noted that these "rare," "unusual," and "freak" accidents have not, heretofore, been taken as evidence that stingrays are capable of something strategic—like *aim*?

⤝⤞

The Glory of the Sea

COMMON NAME: Glory of the Sea Cone	
SCIENTIFIC NAME: *Conus gloriamaris*	
SPECIES RANGE: Indo-Pacific Ocean	

There is a brisk trade in seashells, particularly those that are decorative or rare. Consistently valuable are the cowries. With a seemingly infinite variety of speckling, mottling, and striping, these durable marine gastropod shells are pleasantly smooth to the touch. Their lacquered appearance is the result of wrapping their mantle tissue around the outside of their shells while still living. Long before the electronic cryptocurrency bitcoin, the rather drab-looking cowrie (*Cypraea moneta*) was a state-independent coinage for thousands of years, from the kingdom of Kongo to Hong Kong. French naval officer and explorer Jean-François de Galaup, comte de La Pérouse (1741 to c. 1788), encountered as much in his own ill-fated global circumnavigation. On the heels of stinging victories over the British navy during the American Revolutionary War, the Treaty of Paris freed up King Louis XIV's navy for more expeditionary pursuits. Trading with cowries all the way, La Pérouse made it as far as New Caledonia, but was never seen again (maybe he ate the pufferfish (see page 72)! The Sun King's patronage had already seen to a transformation of French decorative arts from baroque to the florid and gaudy rococo (in the style of rock and

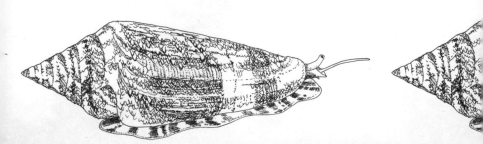

shell gardens). Meanwhile, ships of the Dutch East India Company were returning to Europe from the South Pacific with never-before-seen wondrous and ornate seashells. These fetched handsome sums in northern Europe, where in the homes of the well-to-do they graced Wunderkammer: rooms replete with curio and curiosity, most of which would later serve to seed Europe's natural history museums.

The more rare the shell, the more it would incite the avarice of avid collectors. A hundred and fifty years after the first recorded economic bubble devastated so many ordinary and wealthy Netherlanders with the collapse of the Amsterdam tulip market, the Dutch were seized anew, but this time with conchylomania. The rarest of all shells, however, was not a conch, but the cone snail *Conus gloriamaris*, the Glory of the Sea, with its intricate patterns and high spire of symmetrical whorls. Lawyer and artist-turned-naturalist Pierre Lyonnet (1708–1789), also a noted cryptographer, managed to acquire a

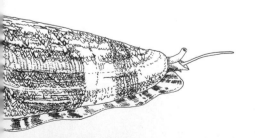

Conus gloriamaris for himself. At the time, he paid well more than three times what Vermeer's *Woman in Blue Reading a Letter* brought at auction only a few years later.

Highly coveted seashells can retain their value as long as their scarcity is undiminished. A pair of rare cowries was even stolen from the American Museum of Natural History in the 1970s. The FBI tracked down the stolen goods in Indonesia, and only a few years later, a Russian trawler proved the folly of the theft when it stumbled upon a glut of the same species near Mozambique. The Glory of the Sea remained one of the most sought after and valuable shells for almost two hundred years.

Not long after Lyonnet's purchase, another Dutch collector promptly smashed the one he purchased at auction with the sole purpose of enhancing the value of his sole other shell. Scads of *Conus gloriamaris* were discovered near Guadalcanal in the 1970s, and the Glory of the Sea presently goes for about $25 apiece on eBay—meanwhile, *Woman in Blue*, one of but thirty Vermeers in existence, sits in its gilded rococo frame in Amsterdam's Rijksmuseum, where it is insured for tens of millions of dollars.

In the end, cone snails may actually be more valuable for their toxins. Ziconitide (valued at $625 per microgram), a recently approved analgesic made from the neurotoxic venom of the magical cone snail (*Conus magus*), is one thousand times more potent than morphine. The various conotoxins all are similar in being very short peptides. These givers and takers of life are aptly labeled, from alpha through omega. Each one hits a different neurological target. Unlike morphine and other opiates, the calcium channel blocker in ziconitide—omega-conotoxin—does not appear to result in tolerance or dependence. There are some side effects to worry about, many of which were already known from unfortunate encounters with cone snails in the wild.

Cone snails deliver their toxic stings with a hardened harpoon. While it can be—and is occasionally—used defensively, the venom of snails in the families Conidae and Terebridae primarily are used to disable prey, which can include fish many times their size. Some cone snails might even begin the incapacitation of their quarry by pumping toxins into the surrounding water before they launch the barbed spear and inject a lethal dose. The speed with which cone snail venom kills is grossly exaggerated. In some circles, *Conus geographus* is dubbed the "cigarette cone"—indulging in one last smoke is claimed to be about as long as you'd have before succumbing to its effects. *Conus textile* was a seemingly efficient murder weapon in the "Cloth of Gold" episode of *Hawaii-Five-O*. In the film *The Lost World: Jurassic Park,* Eddie Carr, who provided the dart that would eventually take down *T. rex* (the dinosaur, not the leech), claimed that the venom of *Conus purpuratus* acts faster than a nerve impulse. Clearly this is ludicrous. Nerve impulses travel at 225 mph—in contrast, to get from skin to brain, injected toxins would have to rely on the slow 2-mph speed of blood through the body or the even slower retrograde axon transport.

Cone snail stings are certainly worrisome and sometimes fatal. There is no antivenom. The neurological effects first include numbness and tingling but can ultimately lead to paralysis and respiratory failure. Since first drawing the attention of Seychellian researchers in the 1930s, there have been a few dozen fatalities, and these only from particularly severe doses. Like snake bites, cone snail envenomations of humans invariably result from intentionally picking up a live snail, usually *Conus textile* or *Conus geographus*, the latter of which is responsible for at least half of human fatalities.

Any species can deliver a blow. In 1935, twenty-seven-year-old Charles Garbutt picked up one of these pretty snails on the Great Barrier Reef. All was fine until he unwisely began to scrape at the external mantle tissue with a knife so as to reveal the smooth, distinctive *Conus geographus* patterning beneath. The snail was having none of it, and thwapped poor Charles in the palm of the hand that held it. After twenty minutes, his vision was blurred. Within thirty, his legs would no longer respond. In another half hour, Garbutt was in a deep coma from which, sadly, he would never recover.

<< >>

Jelly Belly

Anyone snorkeling or diving in tropical waters in the last twenty years has seen them with increasing frequency. Translucent white, maybe blue, or almost clear, and visible only when the sun's rays refract in just the right way. An otherwise glorious afternoon of enjoying the splendor of crystal-clear waters and pristine coral reefs can be ruined by these beastly ghosts of the oceans. People typically see them when it is too late, as they are floating an arm's length away. Each is an unwelcome sight, and there is an enormous diversity of them, to be sure. They range in size from those that could fit in

COMMON NAME: Portuguese man-of-war

SCIENTIFIC NAME: *Physalia physalis*

SPECIES RANGE: Oceans globally

the palm of your hand to others big enough to fill a shopping cart. I am referring not, as you might have been led to believe, to jellyfish and their stinging kin, but rather to that perfidious perennial pestilence of the seas: the plastic bag.

The mendacity of the plastic bag at sea, in many respects, relates to its aptly mimicking scyphozoan true jellyfish and man-of-war siphonophores, both of which similarly float aimlessly adrift. All of that pulsing of the bell of a jellyfish is not directional swimming, except in the case of the box jelly sea wasps (which are actually cubozoans, not scyphozoans [see page 26]). Jellies have no more control over their destination than does a plastic bag tossed overboard by a coed on a pleasure cruise. A jelly's flexing of its bell muscles serves the singular purpose of creating flows and eddies of water over its tentacles so that planktonic prey might be swept into them and snagged by stinging cells. For that matter, no one has ever been attacked by a Portuguese man-of-war. This cnidarian is destined only to be pushed along with prevailing winds by virtue of its floating nitrogen-filled pneumatophore—it lacks any individuated locomotion through which it might "attack."

I am including jellies and siphonophores together only because they are among those cnidarians most frequently annoying to oceangoing humans, be they swimmers, waders, or divers. If you have never been stung, you are simply not spending enough time in the ocean, or perhaps not enough time barefoot on the beach. The stinging cells of the tentacles long outlive the beaching of their larger bodies. In any case, jellies and the men-of-war are somewhat unrelated. Siphonophores (the men-of-war and relatives) have a life cycle that is entirely pelagic, which means that they never touch bottom. The average jelly spends a good deal of time as a small treelike hydroid stuck to a rock or a piece of coral. Further, siphonophore bluebottles mature in a remarkable manner.

The Portuguese man-of-war is a strange sort of colony made up of multitudinous individuals (called zooids): one for floating, some with tentacles for catching prey, others for eating and digestion, and still others for reproduction. However, these are not like free-swimming cyborg "zooids" that are being assimilated from individual to colony. Not one of these individual zooids can live on its own when separated from the "mother ship," nor can any survive without the others. The zooid collective is formed not by differentiation of single cells such that one cell makes two, then those make four, as happens with our own bodies—instead, each new zooid forms by budding from another preexisting zooid. Each is genetically identical and interdependent with the others.

All cnidarians, be they jellies, hydroids, anemones, corals, or even men-of-war, are armed with cells that fire toxic "harpoons" when triggered. The triggering is passive; there is no malice or intent on the part of the jelly. And the toxins, at least to humans, are almost universally simply (if strongly) inflammatory, short-lived, and of little consequence. This is not to say they cannot ruin a snorkeling trip or an otherwise refreshing dip; the incessant electric zaps have ruined plenty an outing. Still, the only cases of fatalities from Portuguese man-of-war stings are anaphylactic immune responses as a result of encounters with a "bloom" of several dozen at a time, each equipped with meters-long tentacles. And in every case, the siphonophores were just minding their own business until the "victim" failed to notice their painfully obvious bright blue and red and magenta hues. (As an aside, I might note the utility of prescription lenses in a mask while snorkeling.)

Not every animal seems to be bothered by jelly tendrils or siphonophore tentacles. Several fish species appear to be immune, and sea turtles are considerably more endangered from hunting and the destruction of nesting habitat by humans than they are from the various jellyfish and bluebottles—species that they devour with aplomb.

All of which brings us back to the real scourge: plastic bags. It would be hard to argue that turtles are simply dumb, as they have been around for almost a quarter of a billion years, outliving dinosaurs and dodos alike. Nor would it be fair to blame tortugan eyesight, except perhaps for its keenness. A sea turtle's love of jellyfish knows no bounds. Alas, this means that turtles evolved to eat things that look a lot like plastic bags. Consequently, sea turtles eat plastic bags with the same voraciousness. They will rush at a bright blue grocery bag mimicking a man-of-war floating on the surface, and they will scarf down a scyphozoanlike zip-top sandwich bag to boot. As the plastic accumulates and is unable to pass through a turtle's digestive tract, it leads to malabsorption, malnutrition, and ultimately death. The shame is ours.

<< >>

First Encounters

COMMON NAME: Asian giant hornet; yak-killer	
SCIENTIFIC NAME: *Vespa mandarinia*	
SPECIES RANGE: East Asia and Japan	

Memory fades with age. Without periodic reinforcement, faces, names, and events from the past gradually dissipate and are lost. I cannot remember the name of my fourth-grade teacher, but I *do* remember, as I'm sure many of us do, every hymenopteran sting

I have ever received—frequent reinforcement does that. Having just been shown, as a six-year-old, how to pluck clover flowers and suck out the sweet nectar from the base of the petals, I was busily engaged in this new delight when a bee thrust its weapon into the palm of my left hand. This was the manner, in which, at six years old, I learned that bees like nectar as much as they are fond of pollen. Bursting through the front door of our home and screaming in pain, my attentive mother removed the stinger and explained that the bee would now die, its guts having been torn out and its stinger left embedded in my skin. This was the manner in which my mother taught me about schadenfreude.

My first encounter with a hornet sting was when I bit into a green apple on a camping trip with the Boy Scouts. Having taken a few bites, I had put the apple down to hoist a pack on my back. My next chomp on the apple was instantly followed by a buzzing in my mouth and a series of sharply painful stings on my cheek and my tongue. And, so, at ten years old, I discovered an important fact about wasps and hornets: they do not lose their stingers and can strike repeatedly. To this day, I am still reflexively wary of green apples.

My first encounter with a yellow jacket was in the 1970s. Child labor laws did not apply to paper routes or farmwork, so, naturally, I did both. My first full-time summer job was on a strawberry farm. Reaching through the straw bed for a particularly plump berry, my left arm was suddenly swarmed by *Vespula maculifrons*, and I received a half-dozen stings. The foreman made a point of humiliating me by insisting they were simply mosquito bites. Thus, at twelve years old, I discovered: that yellow jackets nest in the ground, that bees get wrongly blamed for most hornet stings, and that adults often just try to look smart by making stuff up for their own convenience. It would be another eight years before my next hymenopteran hit when waving away yet another yellow jacket—probably the invasive German species (*Vespula germanica*)—I was nailed on the back of my right hand. Within twenty-four hours, my hand looked like a grapefruit, as the swelling progressed to my shoulder. And so, at twenty years old, I learned that I had developed an allergy to wasp stings. Paper wasps (*Polistes exclamans*) have twice since caused reactions so severe that I was at risk of developing cellulitis in my legs.

The stinger of a wasp is actually a modified ovipositor. Only the queen actually gets to use hers for the original purpose: laying eggs. The vast majority of these eusocial insects are female, and they are destined to only toil for the collective as nest builders, egg tenders, food gathers, and—almost universally—together as a well-coordinated and heavily armed battalion. In most cases, wasp and hornet stings are just painful and inflammatory. The pain is mostly due to a hefty dose of serotonin in the venom. The inflammation results from the other soupy protein components of the mix. It is thought that the mast cells in our skin that react to insect stings are a specific evolutionary adaptation, which allows mammals to locally detoxify hymenopteran venoms. Without them, our ancestors may not have had the benefit of honey to sweeten their short lives. Allergy to the proteins after repeated stings can cause a predisposition to anaphylactic response when stung. However, there are few hymenopteran species that are truly dangerous, with some exceptions, of course. Every year, a few dozen people are killed by the Japanese subspecies of the Asian yak-killer hornet (*Vespa mandarinia japonica*). With a three-inch wingspan and a quarter-inch stinger, this bee-eating hornet is capable of delivering a huge toxic dose of mandaratoxin, a peptide that rapidly blocks all sodium-related nerve cell communication.

Luckily, I have never gone full-blown anaphylactic from wasps or hornets—not yet, at least. But, as a field biologist, I take no chances, and carry two epinephrine syringes with me wherever I go. The anaphylactic allergic response is an unnecessarily body-wide systemic histamine dump reaction to what is really just a local insult. The resultant swelling of the throat and spasms in the trachea can lead to suffocation. Dilated blood vessels leak their contents into surrounding tissue with an attendant whole-body swelling and precipitous loss of blood pressure as the victim slides into shock. A shot of adrenaline constricts the blood vessels, boosts the heart rate, and quickly reverses the life-threatening symptoms.

Perhaps the most dangerous wasps are the mud daubers (*Sceliphron caementarium*), though they are unlikely to sting—even when harassed. A colony is thought to have nested in the pitot tubes of a 757 flight from the Dominican Republic to Frankfurt in 1996. In flight, the plane's air-speed readings failed, causing Birgenair Flight 301 to stall and plummet into the Atlantic Ocean, killing 13 crew members and 176 passengers.

<< >>

Unlucky Thirteen

> COMMON NAME: Red lionfish
>
> SCIENTIFIC NAME: *Pterois volitans*
>
> SPECIES RANGE: Indo-Pacific Ocean; invasive in Caribbean Sea

Wearing thick gloves, take a red lionfish (*Pterois volitans*) and carefully trim off the spiny pectoral fins and dorsal spines—discard them. The scales are easy to remove under running water. The venom glands at the base are toxic only if injected, so the fillets are fine to eat—although a bit small. Next, cut the fillets off each side, remove the skin as with any other fish, pat dry, and deep fry until golden brown. Then enjoy these Caribbean lionfish fritters, but enjoy even more your satisfaction in ridding the Caribbean Sea of a dangerous, invasive species.

Carl Hiaasen's fictional book *Stormy Weather* hilariously captures the many exotic animals that roamed South Florida after the devastation of Hurricane Andrew in 1992. He was not far from the truth. In addition to the thousands of monkeys, parrots, wallabies—and even a mountain lion—that escaped their cages in the aftermath, at least six exotic red lionfish were seen happily swimming in Biscayne Bay after being swept away from their owner's aquarium. Now, whether this alone was *truly* the cause of the Caribbean's current infestation of *Pterois volitans* is unlikely. Genetic studies point to many separate introductions; nonetheless, the true culprits probably are aquarium enthusiasts who release their unwanted pets into the wild, where they rapaciously wreak havoc on ecosystems in which they do not belong.

As with most wildlife, particularly showy fish are best left alone. With striped bodies and thirteen venomous dorsal spikes, the lionfish's pectoral fins are also armed with more than a dozen long, venomous spines webbed together. When threatened, with their fins extended and mouths agape, lionfish resemble Australian frilled dragon lizards ready to strike. Lionfish are aggressive only to fish that are smaller than themselves. Producing millions of eggs each year, and without any natural predators where they have been introduced, lionfish are probably now permanent residents from the Carolinas to the Lesser Antilles.

Among fish, unfortunately, the look-but-don't-touch admonishment holds true regardless of showiness. The most dangerously venomous fish in fact are nearly impossible to see. Lionfish, scorpion fish, devilfish, and stonefish together comprise the family Scorpaenidae, and while a few scorpion fish, just like the many species of lionfish, are resplendently colored and easy to see, the majority of the fishes in this family are stealthy lie-in-wait predators. Their skins are usually decked out in frills and filaments that cover large, knobby skulls attached to nondescript bodies. Taken together, the effect for a stonefish is to look just like a stone or some other piece of rock or coral that has been resting in place long enough to grow thick with hydroids and epiphytic algae—that is, until one steps on it. So, tread lightly, because the thirteen short, stout, hollow spines running down the length of a stonefish's back are equipped with glands that extrude their venom hypodermically when squished by the sole of a foot.

Scorpaenid stings are common throughout the Indo-Pacific region and result in searing pain and localized swelling. In Australia, stonefish antivenom employment is second only to that for red-backed spider bites. Made from horses, the antivenom is an immunoglobulin soup directed at the various myotoxic, cardiotoxic, and neurotoxic proteins that constitute scorpion fish venom. Stings are usually only a problem if the necrotic wound becomes infected, but a few species are extremely dangerous. In 2010, while conducting a lesson in knee-deep water, a dive instructor on a beach in Okinawa, Japan, received deep punctures from the thirteen dorsal spines of a devilfish (*Inimicus didactylus*). In spite of rapid medical attention, an hour and a half later he was gone.

Preparation of stonefish in Japan is less involved than the recipe provided here, as it is either served raw as okoze sushi, or the whole fish is simply deep-fried. The toxins are heat-labile. Regrettably, okoze sushi is *not* actually on the menu at the Okoze Sushi Restaurant in Russian Hill, San Francisco.

━◄━ ━►━

Stinging Bullets

COMMON NAME: Bullet ant

SCIENTIFIC NAME: *Paraponera clavata*

SPECIES RANGE: Tropical Central and South America

My personal encounters with ants are more frequent and less worrisome than those I have had with wasps. Fire ants (*Solenopsis invicta*) in the Deep South have been a constant annoyance ever since they were introduced from South America. I once had to relocate a television shoot for National Geographic in Venezuela in light of half a million army ants (*Eciton burchelli*) that decided to plow through our carefully chosen understory setup. Antbirds were leading the front, snatching up all of the fleeing cockroaches and millipedes. Later in Argentina, we were ill located in the way of a nocturnal wave of marauding army ants. I didn't bother to identify them as they dropped onto my face from the rafters. On a separate expedition in Zambia and Rwanda, I spent some time dodging massive waves of migratory driver ants (genus *Dorylus*)—luckily, they can sting but they don't always. Their enormous mandibles inflict enough pain from the shearing action they use to tear into flesh. It is commonly believed that when swatted away from the skin, driver ants release alarm pheromones, which cause their fellow comrades to all chomp down simultaneously. On yet another expedition, this time in French Guiana—where even the trees sting—I was able to distinguish the Brazilian red ant (*Solenopsis saevissima*) from the electric ant (*Wasmannia auropunctata*). One felt like a white-hot needle; the other felt like a smoldering cigar being crushed on my foot.

Pain is notoriously difficult to quantify due

to its subjectivity. As an entomologist at the Carl Hayden Bee Research Center in Arizona, Justin Schmidt has made a point of being stung by more Hymenoptera than the average honey badger. In light of his careful, if colorful, note taking, there is now a Schmidt Sting Pain Index. South American bullet ants (*Paraponera clavata*) are fearsome and fearless. Schmidt describes the sting of the bullet ant as "immediate, excruciating pain and numbness to pencil-point pressure, as well as trembling in the form of a totally uncontrollable urge to shake the affected part." His pain scale ranges from the merely annoying level 1, a sweat bee sting he experienced, to level 4, the sting of a tarantula hawk wasp in the genus *Pepsis*. However, the aptly named bullet ant gets a level of 4+ from Schmidt, who elsewhere detailed it as "pure, intense, brilliant pain. Like fire-walking over flaming charcoal with a three-inch rusty nail in your heel."

Sorex araneus

Bungarus multicintus

Crotalus adamanteus

Ixodes holocyclus

Heloderma suspectum

4

THINGS THAT BITE

Bad Bill

COMMON NAME: Gila monster

SCIENTIFIC NAME: *Heloderma suspectum*

SPECIES RANGE: Sonoran Desert

The drive from southern New Mexico on back roads in early July with my friends Lee and Lynn took us below the

Mogollon Mountains and along the Gila River. Among cracked dirt banks with sharp rocks amid a fine layer of dust on low-lying scrub and teddy-bear cactus, I could smell the faint whiff of ozone before hearing, or seeing, any water still hours from Tucson, Arizona. We were in the eastern edge of Gila monster territory—a parched, thirsty land that continues all the way to Yuma. Among other folklore surrounding *Heloderma suspectum*, a Gila monster sighting was a sign of impending rain, according to the Apache; however, it is more likely a sign of rain having already come. These robust and unnecessarily feared lizards spend 95 percent of their lives underground, where temperatures are moderate.

For millennia, indigenous peoples from Oaxaca, Mexico, to the Sonoran Desert have associated Gila monsters more with negative superstitions than with a promise of something as joyful as rain. Just seeing a *cheadagi* (as the Tohono O'odham nation call this lizard) was thought to induce outbreaks of festering sores. That the genus name for the Gila monster, *Heloderma*, begins in a manner homophonic with the Gila (pronounced *Hi-laə*) River is accidental. *Helo* and *derma* refer to the lizard's osteodermal bony skin. In a Kiplingesque "How the Gila Got Its Hela," there is an Aztec myth that describes how one saved a corn crop and thus came to have bony skin resembling corn kernels.

Encounters with Gila monsters are rare, and rarely dangerous. There are as many fictitious accounts of fatalities from their bites as there are fictitious frontier stories of Wild West gunfights in the 1800s. Fear of this lizard may simply relate to its scarcity and its unusual appearance. *Heloderma* species have knobby, orange and black marbled skin, a snakelike forked tongue, a ridiculous-looking fat stumpy tail, and a menacing face. Their tails are a convenient place to store fatty tissue, allowing them to survive through the lean months of the underground summer.

Perhaps the best "documented" death from a *Heloderma* bite was that suffered by a pool-hall proprietor in the 1930s, after a night of boozing and playing with his pet Gila monster. The creature chomped down hard on the proprietor, who was dead within the hour. Of course, the truth is that the bumbling billiards boss was blindingly drunk. He would have died of alcohol poisoning within another hour—lizard laceration or not.

This is not to say that Gila monster venom isn't worrisome. The pair of large glands on either side of its lower jaw beneath the lower lip (if lizards had lips) bulge with a venom as fearsome as some snakes'. Envenomation is powerfully painful and strong enough to kill a small dog if the lizard manages to deliver sufficient quantities when harassed. Even in humans, the venom causes a sharp drop in blood sugar while also inducing bulging eyes. What they lack in size of tooth, Gila monsters make up for in tenacity. There's even an account of a particularly persistent "Bad Bill" clamping itself onto a car-door handle for over fifteen minutes. Their long, hard grinding serves not only to drive venom deeper into tissue, but the force of it helps squeeze the venom glands, allowing them to deliver more of the toxic cocktail than would otherwise be possible.

≺≺ ≻≻

Measure for Mammal

COMMON NAME: Common shrew
SCIENTIFIC NAME: *Sorex araneus*
SPECIES RANGE: Europe and Northern Asia

The shrew has been associated with darkness and danger for thousands of years. For the ancient Egyptians, symbolism both for light and dark, and for seeing and blindness, is strongly associated with the god Horus, who, after suffering indignity upon indignity, and with the help of his mother Isis,

bested his brother Set for the throne of Egypt. In one incident, Horus was blinded by Set. As such, the Eye of Horus remains one of the most recognizable Nilotic symbols to this day. However, Horus takes many forms, not all of which are benevolent. Horus Khenty en Maatyu is a pharaonic symbol of darkness in the form of a shrew roaming the Underworld and torturing those who led evil lives.

The Romans feared shrews as well, thinking it possible for a shrew to take down livestock with a single bite. As a precaution, dead, dried-up shrew bodies were hung like talismans around the necks of cattle, giving far too much credit to the shrews' toxicity and their ability to recognize fellow mummified kin. In truth, shrews aren't nearly so shrewd! The scientific species name for the common shrew, *Sorex araneus,* is actually a reference to spiders. Our name for the common shrew, or shrewmouse, has its origins in the Old English word "screuua" or "scraewa." Whether the Britons were influenced in their beliefs about shrews by the Romans during their period of occupation is unclear. However, by the 1300s, this unassuming little mammal, which nests underground or under leaf litter, was already held in rather low esteem. English bestiary author Edward Topsell (1572–1625) characterized the shrew as "a ravening beast feigning itself gentle and tame . . . being touched it biteth deep and poisoneth deadly." Mind you, this is the same guy who thought weasels gave birth through their ears.

To "beshrew" someone was to invite evil on the one so cursed. Cursed, wicked—i.e., shrew'd—people could, of course, be wily and mischievous in their evil ways; they were shrewish, shrewdly, or just shrewd. Only later did the word eventually lose its negative connotations and evolve to mean a person of enviable cunning. To be shrewish first referred to the cursèd, and then later inverted to the one who does the cursing . . . or scolding. For Chaucer, men and women alike could be labeled "shrewes" if they were taken to scolding and berating others. As the English language advanced into the Elizabethan era, a "shrew" was universally taken to refer to an incorrigible scolding wife. Shakespeare's most misogynist play, *The Taming of the Shrew*, has seen periodic revival in *Kiss Me, Kate*, John Wayne's *McLintock!,* and even Richard Burton and Elizabeth Taylor's art-imitating-life adaptation.

Only a few mammal species have venomous bites, and most that do are closely related, such as the Hispaniolan solenodon (*Solenodon paradoxus*) from Haiti and various shrews in the genera *Sorex*, *Neomys*, and *Blarina*. As far as mammals go, shrews are really quite small. With the metabolism of a hummingbird, they are required to eat about three times their own weight every day, and foraging for food is a daunting task. Shrew venom is delivered from salivary glands and, as with Gila monsters, along grooved teeth. The blarina toxin, or BLTX, is an enzyme that induces uncontrolled activation of the bite victim's own naturally occurring bradykinin. Localized bradykinin activation, whether from a shrew or a kissing bug, is extremely painful to something the size of a human, but the same bite is positively paralytic, if nonlethal, to smaller prey.

So, the next time you encounter a cute little shrew, give some thought to the still-alive grubs that are paralyzed and stacked up in its den like salami in a deli case waiting to be eaten. Soricidin, a salivary peptide secreted by the northern short-tailed shrew, is presently under investigation for biomedical applications by some shrewd Canadian researchers.

<-<- ->->

Red–Green Show

> COMMON NAME: Bloodworm
>
> SCIENTIFIC NAME: *Glycera americana*
>
> SPECIES RANGE: Global

Bloodworms make pretty good bait for catching near-shore marine fish like stripers, flounder, and bluefish (see page 82). There is a lucrative market for these luscious lures, too. Going for up to $2.50 a piece, *Glycera americana* is pound for pound the priciest marine resource in Maine. Now that the bottom has fallen out of the lobster market, it may be the most profitable as well. At its peak, the Maine bloodworm harvest exceeded 600,000 pounds, amounting to nearly $8 million,

making bloodworms one of the top five most important fisheries in the state. *Glycera americana* is a polychaete, which are annelid worms that are distant relatives of other baitlike earthworms and leeches.

Harvesting bloodworms is often a well-guarded affair. Listening to the diggers tramping knee deep in muck, one would think they were in the forests of Périgord, France, hunting truffles. Prized sites are closely guarded secrets that are passed from one generation of vermi-sleuth to the next. Bloodworm diggers are practiced connoisseurs of muck. Too sandy and the worms won't be there; too dense and one cannot get to their burrows. It takes a quick eye and a sturdy back that can withstand hours stooped over, pulling mud with a hand rake at low tide. Old-timers do it all by hand, which can prove painful.

A voracious carnivore, *Glycera americana* grows up to a foot in length and has a fat, fleshy proboscis armed with four wickedly sharp and inward-curving fangs. The bigger the worm, the bigger are the spikes. Each fang has its own poison gland that packs a toxin strong enough to kill other worms, and more than enough to inflict some serious pain on an ungloved hand. The four scythes have a hardness that exceeds the dentin of teeth and are fashioned from a blackish-green crystallized form of copper, atacamite. Other than in the makeup of the fangs of bloodworms, atacamite is found only as a semiprecious stone in the northern arid deserts of Chile. To produce these green spikes as they grow, bloodworms have to put up with a body burden of toxic, concentrated copper that would easily damage the brain or liver of other, lesser beasts like us.

<-<- ->->-

Ticked Off

COMMON NAME:	Australian paralysis tick
SCIENTIFIC NAME:	*Ixodes holocyclus*
SPECIES RANGE:	Eastern Australia

An old Irish proverb holds that "An Irishman is never drunk as long as he can hold onto one blade of grass and not fall off the face of the earth." It would seem that ticks, too, have never had quite enough to drink, for the tips of many a blade of grass are precisely where you will find the thirstiest of these arachnids. Having eight legs (as adults), no doubt, helps in their peculiar "questing" behavior. Holding onto the highest point of a blade of grass, or the farthest edge of a low-lying leaf, with one or more of its hindmost pairs of legs frees the other two pairs to grasp at the hair (or clothing) of a passing mammal. For the myriad of annoyances larger dinosaurs may have faced (like asteroids), ticks were not among them because grasses did not appear until after the last dinosaur was extinct. And, as grasses were spreading across the globe about 60 million years ago, so, too, were ticks . . . just in time for all of those tasty new mammals.

Although ticks may have diversified alongside mammals, they are common enough on snakes, birds, and lizards, even having been found once on a crocodile. While it is true that some ticks typically appear only on certain species of birds, for example, other ticks aren't as picky. The lack of preference for a particular host by many hard ticks (Ixodidae) is what renders them so problematic. Any bloodsucking arachnid that, after hatching, feeds on a mouse; after molting, feeds on a deer; and, after reaching adulthood, feeds on a person or their pet is the perfect vector for spreading nasty infectious agents such as Lyme disease, Rocky Mountain spotted fever, Q fever, relapsing fever, encephalitis viruses, babesiosis, and tularemia, just for starters.

However, it wasn't only disease that made exploration of the Australian Outback in the first half of the nineteenth century such a perilous undertaking. The history books are filled with tales of challenge and hyperbole. Two explorers, Hamilton Hume and William Hovell, left Canberra for a two-month overland hike to Melbourne and returned in November of 1824—a feat on foot that today can be accomplished in under two weeks. Hovell wrote of a "tick, which buries itself in the flesh, and would in the end destroy either man or beast if not removed in time." As unlikely a tale as this would seem, and unnecessarily dramatic in a country already so well known for its venomous snakes and spiders, Hovell's warnings were prescient. Soon, the ample grazing lands found beyond the Great Divide were supporting herds of longhorns and, soon enough, cattle, especially calves, were becoming paralyzed. First the hindquarters would stagger and collapse, then full paralysis and death would follow. In most cases, a single thoroughly engorged tick would be found on the back or the flank.

The notion that a tick no bigger than a kernel of corn could take down an animal more than two hundred thousand times its size was rightly met with some skepticism. But by the 1920s, it was clear that the hard tick *Ixodes holocyclus*, did, indeed, have a venom powerful enough to pull off this feat. Alarmingly, these paralytic ticks were not just killing cattle, horses, and dogs, but children, too. The gravestones of Cooktown, Australia, still speak to the horrors inflicted by such ticks on unsuspecting early settlers.

Today, tick paralysis is known to be a global problem. Over sixty different species of hard ticks and soft ticks have been implicated. The condition afflicts domestic and wild mammals from camels in Somalia to flying foxes in Australia—but never marsupials. By the mid-twentieth century, there were dozens of human cases occurring in western parts of Canada and the United States, with children and women being the most common victims.

The exact nature of the toxins in tick saliva is still being investigated today, but the specifics of tick paralysis are now well understood. Ticks will feed either on a single individual, or up to three, in a lifetime. After burrowing its mouthparts into the flesh of its victim, the tick (especially adult females) can stay attached for days and weeks on end, filling up to more than five hundred times its unfed weight. It is these prolonged feedings that allow injection of enough venom to incapacitate a host.

Paralysis generally starts with weakness and loss of coordination in the legs and arms within five days after the onset of the tick's feeding. The toxin seems to function by blocking the way nerves normally stimulate muscles to work. Eventually, paralysis progresses up the trunk, and death from respiratory failure is likely within a few days of the first signs of weakening. However, while prognosis is poor, treatment is easy: simple removal of the offending tick clears up more than 95 percent of all cases.

◄◄ ►►

Patience

It sounds like the plotline of a Grimm brothers' fairy tale. Patience Mouffet is a sickly girl with a respectable physician for a stepfather who tests his deranged medical theories on the poor girl. When Dr. Mouffet arrives in the parlor one day with another dose of dead spiders for Patience to eat, she finds her courage, stands up to her stepfather, and runs away. Strangely enough, the story of the real Little Miss Muffet is true, at least partly so. Although many of the specifics are not known, Dr. Thomas Mouffet (1553–1604), with a surname often spelled *Muffet*, *Moufet*, or *Moffet* in sixteenth-century London, was a member of the Royal College of Physicians and Surgeons who was intent on being a well-respected scientific author. He expounded on commonsense prevention and healthy lifestyles in his book *Health's Improvement*, the fifteenth chapter of which, not surprisingly, concerns "Of butter cream curds, cheese and whey."

However, it was the first catalog of British insects, *Theatrum insectorum,* that he yearned to see published. The expense of self-publishing at that time typically required substantial patronage either from nobility or from an influential publisher. With the establishment of the Royal College, however, scientific publishing had been transforming into something closer to the peer-review system known today. Alas, for Mouffet, several members of the Royal College were determined to thwart his work. The *Theatrum* would not appear in print until 1658, after having been reworked by Theodore Mayerne about fifty years after Mouffet's death. What impeded Mouffet's success in seeing his work make it to print were his views on toxins and toxic animals: Mouffet was an unapologetic Paracelsian—a follower of the medical movement popular in sixteenth- and seventeenth-century Europe that was based on the theories and therapies of Paracelsus.

Philippus Aureolus Theophrastus Bombastus von Hohenheim (1493–1541) took the name Paracelsus not because of personal disaffection with his otherwise bombastic name, but in fancying himself as having surpassed the knowledge and wisdom of second-century Greek philosopher Celsus—an odd choice since

Celsus is best known for his blistering attack on Christianity's spread throughout the Roman Empire. The teenaged Paracelsus worked as an apprentice in the mines of Austria, where he delighted in mineralogy and metallurgy while developing a mystical fascination with alchemy. Following in his father's footsteps, Paracelsus earned his medical degree at the University of Ferrara, Italy, in 1515. He showed contempt for the classic teachings of Galen and championed the use of chemistry in place of the more conventional herbal medicines of the day. In Basel, Switzerland, Paracelsus and his ideas exploded on the scene when his methods "cured" the infected leg of the influential publisher Frobenius. When he had the temerity to boast to local medical authorities that "every little hair on my neck knows more than you . . . and my beard has more experience than all of your high colleges," Paracelsus was banished from Basel. Paracelsus gave us zinc, laudanum, and the phrase "the dose makes the poison," presumably referring to toxicology, and not to his unrestrained braggadocio. He spent the next dozen years of his short life wandering Europe in search of occult knowledge, writing about cosmology and surgery alike.

On his death, Paracelsus's legacy was a divisive one in its rejection of conventional wisdom, particularly in medicine. Thomas Mouffet himself suffered the wrath of the medical establishment and was forced to redact sections of his doctoral dissertation in light of their patently Paracelsian character. Mouffet held firm to the notion that no single thing was entirely poisonous, but that when used in small and appropriate quantities, even venomous spiders could work wonders in healing any and all ailments. With wealthy patrons backing Mouffet, the Royal College could not overtly block his admission, but they could certainly thwart the publication of his unfounded conjectures. Mouffet was unique in invoking the power of spiders—they do not even show up in the witches' brew recipe of his contemporary, William Shakespeare.

Spiders are sadly misunderstood and often needlessly feared creatures. Their venom can be powerful stuff. It has to be, in order to quickly immobilize an insect caught in their diaphanous silk, or to subdue something larger than the spiders themselves. The largest spiders are the tarantulas, but these are among the least worrisome. One is much more likely to get a rash from the tiny, irritating hairs they fling around than from a tarantula bite. One of the most feared spiders is the Sydney funnel-web of Australia (*Atrax robustus*). Equipped with fearsome long, thick, and sharp fangs, it can overcome even lizards and frogs that happen to trigger the silky trip lines radiating out from a burrow at night. But for all of the hype, there hasn't been a single death from a funnel-web spider bite in over thirty years. Only 10 percent of bites are serious enough to treat with widely available antivenom that counteracts what is typically a slow-acting poison. In the time it takes to dance the Italian Tarantella a few times there should be full recovery. What is odd about the effects of the Sydney funnel-web spider is that local mammals—like the marsupials in Australia—appear to be entirely immune to its neurotoxicity. In Australia, humans are the only susceptible mammals.

Atrax robustus and various neotropical species of the genus *Phoneutria* exhibit similar aggressive postures when they are threatened, in which they

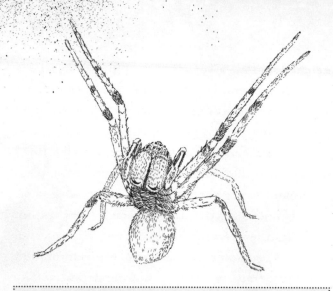

COMMON NAME: Brazilian wandering spider

SCIENTIFIC NAME: *Phoneutria nigriventer*

SPECIES RANGE: Brazil, Northern Argentina

stretch their front-most pair of eight long, hairy legs skyward, bare their fangs, and prominently display a red or orange warning patch. Commonly known as the Brazilian wandering spider, or the banana spider, there are at least eight species of *Phoneutria* found from Costa Rica to Paraguay. In November of 2013, Consi Taylor of London was already a few bites into a Colombian Fairtrade banana when, much to her horror, spots on the banana skin proved to be hundreds of recently hatched spiders, all of which promptly scurried off to the cracks and crevices of her Hampton home. Wandering spiders are extremely aggressive and potently venomous, but even species of *Phoneutria* are only rarely deadly to people. Of the 1,857 wandering spider bites documented over two years in Brazil during the late 1980s, none were fatal. This is not to say that these species' bites are not painful; they certainly are. Worse, the injected toxins cause some serious tissue inflammation that can persist for a week. But the defensive bite

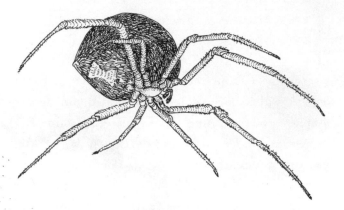

COMMON NAME: Southern black widow

SCIENTIFIC NAME: *Latrodectus mactans*

SPECIES RANGE: Southeastern United States

of any spider is one that can often be dry. The choice to inflict a venomless bite conserves the precious liquid while still inflicting sufficient sharp and sudden pain to make a large predator think twice.

We know quite a bit about spider venoms and their effects as a result of the curiosity of a few arachnologists. Even as conquistador Hernán Cortés was busy wiping out the Aztecs in Tenochtitlán, Mexico, historian Gonzalo Fernández de Oviedo y Valdés (1478–1557) was busy in what is now northern Colombia marveling at "spiders of marvelous bigness." Valdés's *Historia general y natural de las Indias* (1526) was the first written account of the natural wonders of the New World, with tarantulas and bird-eating spiders included. Fearsome looking and enormous in size, tarantulas were widely believed to be deadly venomous until entomologist William J. Baerg (1885–1980) proved otherwise. His intent was to test the bites of various species on lab rats. When he picked up a particularly pugnacious *Avicularia velutina* from Trinidad, it harmlessly bit one of Baerg's fingers. Intrigued, he induced a Panamanian *Sericopelma communis* to drive its fangs into a different finger.

And though this time it hurt quite a bit, the swelling dissipated over the next day or so. This wasn't the first time Baerg had resorted to self-experimentation. Thirty years earlier, finding a first black widow to be too reluctant to bite him, he picked up a second one. To Baerg's delight—initially, anyway—she promptly plunged her fangs into the third finger of his left hand, where she was allowed to stay pumping venom for five or six seconds. In light of what happened next, it is surprising that Baerg would ever allow a spider to bite him again. Within an hour, the pain had radiated all the way to his left shoulder. At two hours, his diaphragm was in spasm, making breathing labored and speech impossible. Seven hours later, a nauseous Baerg was hospitalized as the pain coursed through his whole body. It took several days of hot baths to fully recover.

Dr. William Blair performed the same stunt a decade later when he made careful recordings of his torturous personal journey with the venom of a Southern black widow (*Latrodectus mactans*). Much as Baerg noted, one of the telltale signs was abdominal rigidity from a spasmodic diaphragm. Doubled over with pain, in a fetal

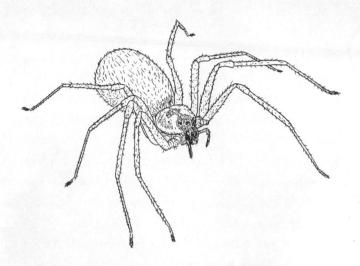

COMMON NAME:	Brown recluse
SCIENTIFIC NAME:	*Loxosceles reclusa*
SPECIES RANGE:	Midwestern United States and south to Texas

position, Blair's blood pressure dropped to 75 over "too faint to determine." The next three days of hospitalization were fraught with agony and uncontrollable tremors, and his skin itched for another two weeks.

As with black widows, it is right to fear brown recluse spiders (*Loxosceles reclusa*). The good news is that while they do like backyard sheds and the debris surrounding places of human habitation, they do not like humans very much. The recluse, or violin, spiders of the genus *Loxosceles* along with their assassin spider relatives in the genus *Sicarius* are among the most medically important in the world. Envenomating bites range from mere inconvenience to necrotizing toxins, which leave festering sores that won't heal. Serious scarring can result. Gangrenous infections that are not properly seen to can result in amputation.

In his book *Dancing Naked in the Mind Field*, Nobel Prize–winning biochemist Kary Mullis claimed he was bitten by a brown recluse—*and* that it came back for more, *and* that it laid eggs, *and* that he was visited by a glowing green holographic raccoon. However, no sicariid spiders occur in the part of California where this was to have occurred, leaving the story about as believable as multidimensional physics on a macroscopic scale regarding the raccoon.

Entomologist Richard Vetter of the University of California–Riverside crowd-sourced spiders in 2005 over the Internet, promising to identify each one for anyone who sent them. Of the 581 spiders that Californians *believed* to be brown recluses, only one proved to be so.

The Arizona recluse (*Loxosceles arizonica*) has a very restricted distribution in the southeastern part of the Golden Gate State. In contrast, the brown recluse is very well known to Missourians and is usually properly identified. The very worst spider species may be the Chilean brown recluse (*Loxosceles laeta*), which I gingerly removed from my sleeping bag while on expedition in the Island of Chiloé in the Pacific Ocean; a good thing, too, as the venom of this recluse has a particularly nasty dermonecrotic enzyme more commonly seen in flesh-eating bacteria. While most recluse wounds are only locally problematic, both *Loxosceles laeta* and African species of *Sicarius* can inject enough venom to cause systemic reactions when the toxins are carried to internal organs.

Thomas Mouffet had, shall we say, some rather uncommon powers of observation. He asserted that people living in houses full of cobwebs and breeding spiders were uniformly gout free, and that "running of the eyes is stopped with the dung and urine of a house spider dropped in with Oyl of Roses." None of this is true of course, but spiders actually are medically useful. The Californian funnel-web spider (*Hololena curta*) produces HF-7 neurotoxin, which is being explored as a way to limit nerve cell death in stroke victims. GsMTx-4 from the venom of the Chilean rose tarantula (*Grammostola spatulata*) might prevent heart fibrillation and is being tested in the treatment of muscular dystrophy. And, insofar as Brazilian wandering spider bites can induce priapism, perhaps the next blockbuster drug will be its PnTx2-6.

◅ ▻

Don't Tread on Me

O wing to the good sense of most snakes, and rather contrary to Theodore Roosevelt's famous pronouncement about the merits of walking softly, it is a heavy foot that serves best in avoiding snakebites. Mind you, having a big stick isn't a bad idea.

COMMON NAME: Puff adder

SCIENTIFIC NAME: *Bitis arietans*

SPECIES RANGE: Africa

Although quite deaf in the usual sense of the word, ophidians of all sorts, whether lounging in reverie in the sunlight or skulking about the base of a bamboo tree, are equipped with keen inner ears with which they readily detect a myriad of vibrations and pressure changes. As such, the average snake will have sensed you and scurried off long before you are within striking range.

Alas, adders are not your average snakes. Stout and slow to move, Africa's puff adder

(*Bitis arietans*) relies on camouflage, whether hiding from predators or lying in wait for unsuspecting rodents—after which it strikes with blazing speed. Adders are indolent reptiles, slow to agitate but unlikely to get out of your way. It is said that it will be the second person in a line perambulating on the savannah who will get nailed by the adder, the first having already annoyed the lurking snake.

Ozzies like to point out that of the twenty-five most venomous snakes, twenty-one occur in Australia. This might lead one to believe that Australian snakes and snakebites are more dangerous, but when it comes to snakebites and deaths from snakebites, Australia barely registers. Ozzies suffer about the same number of envenomations as do the people of tiny Burkina Faso, a country with less than 2 percent of Australia's geographic area. Burkina Faso, however, has the world's worst snakebite mortality rate, with death resulting from almost 20 percent of all cases. Most of these are from puff adders. India is not far behind Burkina Faso, with a one-in-seven chance of death from envenomation. India also has far and away the highest total number of snakebites, seventy-five times as many as occur in Australia, adding up to more than eighty thousand snakebites each year from cobras, saw-scaled vipers in the genus *Echis,* and, the worst of all, Russell's viper (*Daboia russelli*).

Less populous Sri Lanka, with similar species, comes in second after India at thirty-three thousand bites for a whopping one bite for every five hundred people each year. In terms of bites per capita, Sri Lanka is right behind less populous Bhutan, where one out of every three hundred people in the kingdom are envenomated each year and children carry bottles of alcohol to and from school for the purpose of dumping on the ubiquitous snakes including the nabue, lungtham, kasha bue, jagpa bue, and rubjamsa. Thankfully, Bhutan is on par with Australia in terms of a low likelihood of death if bitten. All of this goes to show that danger is entirely relative.

The Australian inland taipan (*Oxyuranus microlepidotus*) has the most lethal venom of any snake on land. The Australian eastern brown snake (*Pseudonaja textilis*) comes in second for power of venom, and the coastal taipan (*Oxyuranus scutellatus*) is third. The taipotoxins in taipans and the textilotoxin in brown snake venom are effective neurotoxins even in minute doses. Their neurological target is the most ancient and universal neurotransmitter across all animal life: acetylcholine. Skeletal muscles and smooth muscles like the diaphragm, heart muscles, central nervous system, peripheral nervous system, and brain stem all use acetylcholine. Blockage of acetylcholine nerve transmission spells death—a paralytic, asphyxiating, heart-stopping death. Presently, there is an antivenom available for Australian taipan and brown snake bites, as well as for black snakes, sea snakes, and death adders (which are not actually adders). However, this wasn't always the case.

In postwar Australia, herpetologists Neville Goddard, Roy Mackay, and Kevin Budden spent a good deal of time looking for venomous snakes, especially in Queensland. Their objective, like that of many others at the time, was to create lifesaving antivenoms against the most dangerous snakes. The concept wasn't new, and already the methods had been perfected in the Butantan Institute of São Paulo, Brazil, in which venom was milked from the snake's fangs, diluted, and repeatedly injected into a horse or sheep until the animal was immune to the proteinaceous toxins. Afterward, it was just a matter of taking pints of blood and separating the serum from blood cells and clotting factors. In principal, this can be accomplished for any animal's venom, and has been.

In 1950, Kevin Budden was intent on finding a taipan for exactly this purpose. Thankfully for the farming families on the Atherton Tablelands, taipans are not common snakes. However, when they are encountered, taipans either rapidly flee or are beaten to death with farm implements. Because a sheep would have to be inoculated several times over the course of several weeks, a dead snake was no good. Budden and Roy Mackay had to find one alive and healthy. This was proving difficult. One morning, Budden was searching for snakes alone by the Flecker Botanical Gardens in Cairns, Queensland, when he heard the unmistakable screech of a rodent becoming a snake's breakfast. Following the sounds, and turning over a piece of discarded Sheetrock, there it was: a large copper-brown, beautiful *Oxyuranus scutellatus*. Being a trained herpetologist, Budden expertly put his foot down on the taipan's neck, causing it to release its already dead prey, and then bent down and grabbed the snake behind the head. The snake, in its desperate—if futile—attempt to escape, thrashed and coiled around Budden's arm so fiercely that he needed both hands to keep it under control. Without a third hand, the snake bag wasn't going to be useful. He was going to have to carry it out and just not lose his grip. It was going to be more than an hour's tramp back to the Cairns Museum. Jim Harris, an observant passing driver on the road, took note of this young man struggling with a snake and stopped, backed up, and offered Budden a ride. Harris was rightfully distressed when Budden explained, as he climbed into the cab, that he was holding one of the world's deadliest snakes in his progressively aching hand. By the time it could be bagged, the snake had drooled so much spit from its gaping mouth, and Budden's hand had become so cramped, that the snake writhed free and, with lightning speed, struck Budden twice on the left hand. Although the snake succeeded in freeing itself, Budden quickly recovered from his surprise and was raced to the local hospital, but not before he recaptured the snake, properly bagging it, and insisting that it be kept safe from harm—it was just too valuable for the lifesaving antivenom that could be produced. A little more than a day later, in the afternoon of July 28, with nothing more to be done and no antivenom to be administered, Kevin Clifford Budden died. He was twenty years of age.

Shipping and Handling

With the exception of puff adders in Africa and Russell's vipers in India, most snakebites occur from picking up snakes. Kevin Budden wasn't the first herpetologist to perish from a passion for snakes, nor will he be the last. The antivenin program pioneered by the Butantan Institute was a joint commercial venture of Mulford Biological Labs in Pennsylvania, Harvard Medical School, and the United Fruit Company. The latter was and remains widely known for its Chiquita bananas distributed worldwide, most of which are grown on Central American plantations. The United Fruit Company had established a serpentarium in the beautiful seaside town of Tela located on the Caribbean coast of Honduras, with Douglas March as its director. March grew up in Haddon Heights, New Jersey, in the roaring twenties. The United Fruit Company's interest was due entirely to their Honduran workers being terrified of snakes on the banana plantations. Most feared of all was the barba amarilla, or yellow beard, (*Bothrops asper*). As plantations were expanding, this very dangerous species was leaving the dense forests, preferring the relative open understory of banana plantations for their nocturnal hunting of increasingly plentiful rodents. Young and ambitious, March had been bitten over a dozen times by rattlesnakes, copperheads, various vipers, and even a fer-de-lance (*Bothrops atrox*) before he opened a second serpentarium in Panama City in the early 1930s.

Snake handling is a dangerous business, and it eventually caught up with him in 1939. March died in April. It was from the bite of a yellow beard, a snake less predictable in its behavior than its cousin, the fer-de-lance. Its venom is a complex mix of enzymes including metalloproteases, phopholipases, and serine proteinases, which collectively liquefy flesh, first at the site of the bite and then quickly progressing to encompass vast swaths of muscle as rapidly disintegrating blood vessels spread the venom. Hemorrhagic lectins and disintegrins render the bite victim incapable of clotting. The resulting tissue liquefaction, blood loss, and sepsis is horrific.

My friend Joe Slowinsky landed his dream job—a curatorial position at the California Academy of Sciences—about a year after I last saw him on New Year's Eve in New Orleans—I made the jambalaya, Joe brought the beer. Soon Joe was off to southern China and Myanmar collecting snakes. Our paths never crossed again before September 11, 2001. Unknown to me as I watched the towers come down in New York, somewhere in the darkening night of a Burmese jungle, Joe was struggling for his life. He had reached into a bag to retrieve a small snake, only to be bitten before recognizing it as a juvenile krait (*Bungarus multicinctus*). Soon the effects of bungarotoxin were shutting his nervous system down. What began with a tingling in his hand that morning proceeded to respiratory failure by noon. After twenty-six hours of heroic

| COMMON NAME: Many banded krait |
| SCIENTIFIC NAME: *Bungarus multicinctus* |
| SPECIES RANGE: Southeast Asia |

efforts to save his life, Joe died from doing
what he loved the most. He's not the only one.
Korean War veteran and herpetologist Fredrick
Shannon (1921–1965) was felled by handling a
Mojave rattlesnake (*Crotalus scutulatus*) in Arizona.
Shannon's close friend and co-author Karl
Schmidt (1890–1957) lost his life to handling

a boomslang (*Bucephalus typus*). Robert Mertens (1894–1975), discoverer of a form of mimicry in which deadly snakes masquerade as harmless ones, was killed from handling a back-fanged twig snake (*Thelotornis capensis*).

Handling venomous snakes is a potentially deadly proposition, even when done by those with years of experience in the pursuit of science. It's alarming what Appalachian Pentecostal ministers do in the pursuit of piety! The biblical passage "Behold, I give unto you power to tread on serpents and scorpions, and over all the power of the enemy: and nothing shall by any means hurt you" (Luke 10:19), might have been referring to the small Levant blunt-nosed viper (*Macrovipera lebetina*) of the Middle East, not the giant eastern diamondback rattlesnake (*Crotalus adamanteus*). Diamondbacks are the most massive of all rattlesnakes and can weigh in at over 30 pounds. Pentecostal minister George Went Hensley, citing the passage from Luke, claimed to have received a revelation from his god commanding all Christians to handle venomous snakes as evidence of their salvation. The practice continues widely today in rural churches of West Virginia, Kentucky, and Tennessee, where the number who have managed to meet their maker is legion. As it happens, Hensley's own temperance preaching didn't keep him from selling moonshine any more than his piety saved him from the diamondback that killed him in 1955.

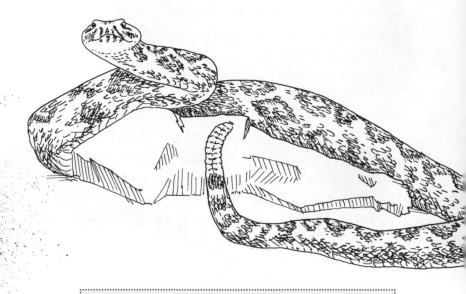

COMMON NAME: Eastern diamondback rattlesnake

SCIENTIFIC NAME: *Crotalus adamanteus*

SPECIES RANGE: Southeastern United States

Snake-handling Pentacostalists are known to be not well acquainted with modern evolutionary theory—and it's killing them. Pit vipers have been locked in an evolutionary Red Queen's Race with opossums for a million years. As opossums have been evolving an ever-stronger resistance to pit viper venom, the snakes have had to evolve more powerful, faster-acting venomosity. Opossums eat snakes and, in most cases, just take a nap when struck by the hypodermic needle–like fangs of a rattler. There is only one pit viper that seems to have won the race: *Bothrops asper* eats opossums. Douglas March was outmatched.

AFTERWORD

Much of what appears in these pages was encountered by the author over the course of two years while researching topics for the exhibition *The Power of Poison* that opened at the American Museum of Natural History in November of 2013. The fascination people have for toxic and venomous organisms is what we tried to capture in the exhibition—without resorting to trivialities like top-ten lists. With the use, for example, of Golden poison frogs by the Emberá of Colombia's Chocó rain forest, as well as the more modern leveraging of a variety of venoms for pharmaceutical powers, history reveals a conflicted relationship with poison. Hidden from view, they are, nonetheless, subject to the same forces of natural selection, evolution, and coevolution as are teeth, talons, and body size. And there are stories, everywhere—some tragic, and some comedic—that richly illustrate the human experience. Shakespeare, the Brothers Grimm, Lewis Carroll, Lemony Snicket, Agatha Christie, and scores of authors have woven poisonous agents into their narratives throughout

history, as did the Greeks and even the Choctaw in their legends. That poisons are unseen actors in the natural world lends a layer of mystery—one that readily captures imagination. Distinct from the exhibition itself, this book has allowed for a different, if idiosyncratic, look at the webs spun through harmful histories and toxic tales. For what it's worth, the author and the illustrator, alike, are optimistic that those reading *this* book will be inspired to pick up another.

<< >>

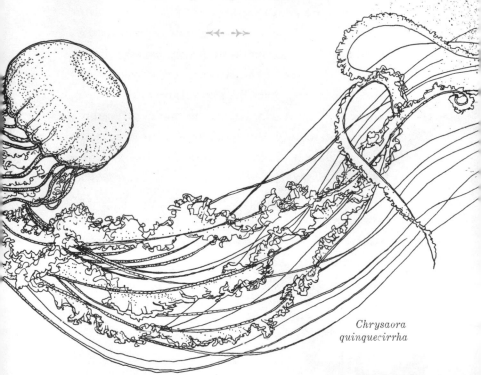

Chrysaora quinquecirrha

ACKNOWLEDGMENTS

The idea of this compendium of stories and anecdotes pertaining to poisonous pests, venomous vermin, and other such artful animals was a result of persuasive encouragements in the year preceding the opening of *The Power of Poison* exhibition at the American Museum of Natural History from Pamela Horn, Sharon Stulberg, and, especially so, from Betina Cochran. Perhaps, more than anyone else, our editor John Foster deserves our greatest thanks. His patience, encouragement, and steady hand, even in the face of the author's own frustrations and tardiness, are deeply appreciated. Facts, anecdotes, and clarifications were offered by quite a few zoologist colleagues, most notable being Darrel Frost, Mande Holford, John Wenzel, Pablo Goloboff, Rich Mooi, Matt Leslie, and R. Joe Brandon. Access at the American Museum of Natural History to specimens for illustration purposes was possible only through the generosity of time and spirit from the Division of Invertebrate Zoology chaired by Jim Carpenter and the Department of Ornithology chaired by Joel Cracraft; in particular we thank Christine Johnson, Paul Sweet, Pamela Horsely, Lou Sorkin, and Sarfraz Lodhi. Capturing the beauty of several animals in a living state was graciously allowed by Fauna NYC and by Jennifer Reed of Red Barn Meadows. Not the least of our thanks go to our close friends and family. The most put-upon by us, and yet never failing to lend an ear to offer candid suggestions and boundless encouragement, were Ali, Mike, and Karen, as well as Michael and Nina.

This book is dedicated to Joe.

⤛⤙ ⤚⤜

GLOSSARY

Below is a condensed glossary of some of the toxins, compounds, and molecules mentioned throughout the essays.

ACETYLCHOLINE: A biological molecule produced by animals that acts as a neurotransmitter for peripheral and central nervous systems.

ALKALOIDS: A class of basic (as opposed to acidic) molecules often (but not exclusively) deployed by plants for their own defense against herbivores; examples include strychnine and morphine.

ANAPHYLAXIS: A very strong immune response, often an allergic one, that is widespread throughout the body; onset is often characterized by hives, followed by swelling and wheezing, which can lead to suffocation and cardiac arrest if not treated.

ANTIVENOM: An antibody produced against a venom, usually purified from the blood of large mammals like horses or sheep that have been intentionally envenomated.

BATRACHOTOXINS: Neurotoxic and cardiotoxic steroidal alkaloids in the skin of the golden poison frog and in the skin and feathers of the hooded pitohui; thought to be concentrated from soft-winged flower beetles eaten by these animals.

BIOACCUMULATION: The process by which a toxin becomes more and more concentrated in tissues of animals that are higher up the food chain.

BRADYKININ: A biological molecule produced by animals that is involved in regulating blood pressure, inflammation, and some pain mechanisms.

BREVETOXINS: Cyclic polyether molecules produced by dinoflagellates that interfere with the normal functioning of nerves by activating voltage-sensitive sodium channels.

CARDIOTOXIC: Toxic to cardiac tissue.

DOMOIC ACID: A neurotoxic heterocyclic amino acid produced by some red algae and some diatoms that causes amnesic shellfish poisoning.

DOPAMINE: A biological molecule that has a wide range of physiological effects depending on where it is acting, such as a neurotransmitter in the brain, a local chemical messenger elsewhere, a vasodilator in the blood stream, or an insulin suppressor in the pancreas.

GLYCOSIDE: A biological molecule that is partly sugar and partly something else; often the something else is a toxin which is activated when separated from the sugar.

HOMOBATRACHOTOXIN: One of several batrachotoxins.

MYOTOXIC: Toxic to muscles.

NEMATOCYSTS: The stinging intracellular organelle in cnidocytes (jellyfish, for example).

NEUROTOXIC: Toxic to neurons.

OKADAIC ACID: A neurotoxic polycyclic fatty acid found in a variety of marine dinoflagellates.

PALYTOXINS: Potent vasoconstrictor toxins from marine organisms that disrupt the ability of cells to maintain a proper balance of sodium outside and potassium inside the cell membrane.

PEPTIDE: A biological molecule made up of a string of amino acids.

POISON: Something that disrupts a normal physiological state or process.

SAXITOXINS: Alkaloid paralytic shellfish toxins produced by marine dinoflagellates.

SEROTONIN: A biological molecule that has a wide range of physiological effects depending on where it is acting: a neutotransmitter in the brain, a local chemical messenger elsewhere, a vasoconstrictor in the blood stream, associated with increased insulin production in the pancreas and pain in the peripheral nervous system.

TETRODOTOXIN: A neurotoxin produced by bacteria (and bioaccumulated by a variety of marine and freshwater animals) that prevents nerves from firing by binding to voltage-gated sodium channels in susceptible nerves (excluding pacemaker cardiac cells).

TOXIN: Typically a specific molecule that disrupts a normal physiological state or process.

VENOM: Something that disrupts a normal physiological state or process if injected.

INDEX